思科系列丛书

思科网络实验室 CCNP（交换技术）实验指南

（第2版）

王隆杰　梁广民　编著

电子工业出版社

Publishing House of Electronics Industry

北京·BEIJING

内 容 简 介

本书旨在帮助正在学习 CCNP 的读者提高 CCNP 交换方面的动手技能。全书共 6 章，主要内容包括交换机基本配置，VLAN、Trunk、VTP 与链路汇聚，STP，VLAN 间路由，高可用性，交换机的安全。本书的重点是实验，希望能通过实验有效地帮助读者掌握技术原理及其使用场合。本书采用 Catalyst 3560 V2 作为硬件平台（IOS 版本为 15.0）。

本书适合想要通过 CCNP 认证考试的网络技术人员，以及那些希望获得实际经验以轻松应付日常工作的专业人员阅读，本书既可以作为思科网络技术学院的实验教材，也可以作为电子和计算机等专业网络集成类课程的教材或者实验指导书，还可以作为相关企业员工的培训教材；同时也是一本不可多得的很有实用价值的技术参考书。

未经许可，不得以任何方式复制或抄袭本书之部分或全部内容。
版权所有，侵权必究。

图书在版编目（CIP）数据

思科网络实验室 CCNP（交换技术）实验指南 / 王隆杰，梁广民编著.—2 版.—北京：电子工业出版社，2018.4
（思科系列丛书）
ISBN 978-7-121-33845-8

Ⅰ.①思… Ⅱ.①王… ②梁… Ⅲ.①计算机网络－信息交换机－实验－指南 Ⅳ.①TN915.05-33

中国版本图书馆 CIP 数据核字(2018)第 048230 号

策划编辑：宋 梅
责任编辑：宋 梅
印　　刷：北京盛通数码印刷有限公司
装　　订：北京盛通数码印刷有限公司
出版发行：电子工业出版社
　　　　　北京市海淀区万寿路 173 信箱　邮编　100036
开　　本：787×1092　1/16　印张：15.25　字数：390 千字
版　　次：2012 年 5 月第 1 版
　　　　　2018 年 4 月第 2 版
印　　次：2025 年 1 月第 14 次印刷
定　　价：68.00 元

凡所购买电子工业出版社图书有缺损问题，请向购买书店调换。若书店售缺，请与本社发行部联系，联系及邮购电话：(010) 88254888，88258888。
质量投诉请发邮件至 zlts@phei.com.cn，盗版侵权举报请发邮件至 dbqq@phei.com.cn。
本书咨询联系方式：mariams@phei.com.cn。

前 言

CCNP 涉及交换的内容不是很多，然而网络工程师在实际工作中和交换机打交道的机会往往比路由器多。作者一直很推崇理论与实验相结合的学习方法，这是作者多年从事教学的经验，也是编写本书的原因。也许理论与工作实践相结合更好，可是又有多少人会有这样的机会呢？会有几个企业允许你在实际的生产环境中实践一下呢？那就让我们先在实验室做实验吧，哪怕你失败也没有关系，实验室中的失败是为了不在实际工程中失败。

本书针对 CCNP 交换部分的考试（代码 300-115）所需的知识精心规划了 37 个实验，这些实验将有助于读者在动手过程中掌握相关的理论。值得一提的是，本书的绝大多数实验可以使用同一网络拓扑实现，这将大大减少读者反复搭建实验台的时间。作者非常期望能通过这些实验帮助读者了解这些技术在什么场合可以使用，将产生什么效果。

全书共 6 章，第 1 章介绍实验台的拓扑以及如何配置访问服务器方便读者进行实验；第 2 章介绍交换的基本配置、VLAN 操作、中继和链路汇聚、VTP，以及私有 VLAN；第 3 章详细介绍各种 STP 技术，包括标准的 STP、PVST、RSTP 和 MSTP，还介绍了 STP 的保护，以及在某些场合可以替代 STP 的 Flex Link 技术；第 4 章介绍使用单臂路由或者三层交换实现 VLAN 间路由的技术，还补充了进程交换、快速交换和 CEF 的区别方面的内容；第 5 章介绍保证局域网高可用性的技术，包括思科私有的 HSRP 和标准化的 VRRP、网关负载均衡和服务器负载均衡，以及如何使用日志服务或 SNMP 监控交换机的运行情况；第 6 章介绍交换机上的各种安全措施，包括基本的端口安全、DHCP 监听、动态 ARP 检测、源 IP 保护、防止 VLAN 跳跃攻击，以及使用 AAA 实现 dot1x 认证和交换机上的各种 ACL。

本书由王隆杰（CCIE#14676 R/S，Security）和梁广民（CCIE#14496 R/S，Security）组织编写及统稿，参加编写的还有刘平、张立涓、石光华、邹润生、石淑华、杨名川和杨旭。编著者虽然已尽全力，书中难免还有错误之处，请发邮件到 wanglongjie@szpt.edu.cn 指正。

编 著 者
2018 年 2 月于深圳

目 录

第1章 交换机基本配置 ··· 1

- 1.1 实验台配置 ··· 1
  - 1.1.1 本书实验台拓扑 ··· 1
  - 1.1.2 访问服务器 ·· 2
- 1.2 实验1：配置访问服务器 ·· 3
- 1.3 实验2：交换机的密码恢复 ··· 8
- 1.4 实验3：交换机的 IOS 恢复 ··· 10
- 1.5 本章小结 ··· 11

第2章 VLAN、Trunk、VTP 与链路聚集 ··· 12

- 2.1 VLAN、Trunk、VTP 与链路聚集概述 ··· 12
  - 2.1.1 交换机工作原理 ··· 12
  - 2.1.2 VLAN 简介 ··· 13
  - 2.1.3 Trunk 简介 ··· 14
  - 2.1.4 DTP 简介 ··· 15
  - 2.1.5 EtherChannel 简介 ·· 15
  - 2.1.6 VTP ··· 16
  - 2.1.7 私有 VLAN ··· 20
- 2.2 实验1：交换机基本配置 ··· 21
- 2.3 实验2：划分 VLAN ··· 26
- 2.4 实验3：Trunk 配置 ··· 31
- 2.5 实验4：DTP 的配置 ·· 35
- 2.6 实验5：EtherChannel 配置 ·· 38
- 2.7 实验6：VTP 配置 ··· 45
- 2.8 实验7：VTP 覆盖 ··· 56
- 2.9 实验8：私有 VLAN ··· 61
- 2.10 本章小结 ··· 65

第3章 STP ·· 66

- 3.1 STP 协议概述 ·· 66
  - 3.1.1 STP（IEEE 802.1d）简介 ··· 66
  - 3.1.2 STP 的加强 ··· 67
  - 3.1.3 PVST+简介 ·· 68
  - 3.1.4 RSTP（IEEE 802.1w）简介 ··· 68
  - 3.1.5 MSTP（IEEE 802.1s）简介 ··· 70

3.1.6　不同 STP 协议的兼容性 ·· 72
　　　3.1.7　STP 防护 ··· 72
　　　3.1.8　FlexLink ··· 74
　3.2　实验 1：STP 和 PVST 配置 ·· 74
　3.3　实验 2：Portfast、Uplinkfast 和 Backbonefast ·· 86
　3.4　实验 3：RSTP ··· 89
　3.5　实验 4：MSTP ·· 92
　3.6　实验 5：STP 树保护 ··· 96
　3.7　实验 6：环路防护 ··· 101
　3.8　实验 7：FlexLink ··· 105
　3.9　本章小结 ··· 109

第 4 章　VLAN 间路由 ··· 110

　4.1　VLAN 间路由概述 ·· 110
　　　4.1.1　使用路由器实现 VLAN 间的通信 ··· 110
　　　4.1.2　单臂路由 ··· 111
　　　4.1.3　三层交换 ··· 111
　　　4.1.4　路由器的三种交换算法 ··· 112
　4.2　实验 1：采用单臂路由实现 VLAN 间路由 ··· 113
　4.3　实验 2：采用三层交换实现 VLAN 间路由 ··· 116
　4.4　实验 3：在三层交换机上配置路由协议 ··· 119
　4.5　实验 4：路由器上的 3 种交换方式 ··· 125
　4.6　本章小结 ··· 132

第 5 章　高可用性 ··· 133

　5.1　高可用性技术简介 ··· 133
　　　5.1.1　HSRP ··· 133
　　　5.1.2　VRRP ··· 135
　　　5.1.3　GLBP ··· 136
　　　5.1.4　SLB ·· 137
　　　5.1.5　Syslog ··· 138
　　　5.1.6　SNMP ··· 139
　　　5.1.7　交换机堆叠 ··· 140
　5.2　实验 1：HSRP ·· 141
　5.3　实验 2：VRRP ·· 146
　5.4　实验 3：GLBP ·· 150
　5.5　实验 4：SLB ·· 160
　5.6　实验 5：Syslog ·· 166
　5.7　实验 6：SNMP ·· 169

5.8	实验 7：堆叠	174
5.9	本章小结	180

第 6 章 交换机的安全 182

6.1	交换机的安全简介	182
	6.1.1 交换机的访问安全	183
	6.1.2 交换机的端口安全	183
	6.1.3 DHCP Snooping——防 DHCP 欺骗	183
	6.1.4 DAI——防 ARP 欺骗	184
	6.1.5 IPSG——防 IP 地址欺骗	184
	6.1.6 VLAN 跳跃攻击	185
	6.1.7 AAA	185
	6.1.8 dot1x	186
	6.1.9 SPAN	187
	6.1.10 RACL、VACL 和 MAC ACL	188
6.2	实验 1：交换机的访问安全	188
6.3	实验 2：交换机端口安全	194
6.4	实验 3：DHCP 欺骗	201
6.5	实验 4：DAI 与 IPSG	206
6.6	实验 5：AAA	212
6.7	实验 6：dot1x	222
6.8	实验 7：SPAN	228
6.9	实验 8：RACL、VACL 和 MAC ACL	232
6.10	本章小结	235

参考文献 236

第1章 交换机基本配置

本章将首先介绍本书中始终要用到的实验台的拓扑,该拓扑能够灵活地把不同数量的路由器和交换机进行组合,组成各种拓扑以满足不同实验的要求。随后将详细介绍如何配置访问服务器,以便同时控制多个路由器或者交换机。最后,本章还将介绍交换机的密码恢复以及 IOS 恢复过程。

1.1 实验台配置

1.1.1 本书实验台拓扑

为了完成本书的各个实验,需要构建不同的拓扑,如果每次都临时进行拓扑的搭建会花费大量的时间。我们设计了一个功能强大的网络拓扑,如图 1-1 和图 1-2 所示(图中不包含访问服务器和它们的连接),本书绝大多数的实验可以使用该拓扑完成;该拓扑还可以满足 CCNA 以及 CCIE 的部分实验。拓扑中的路由器和交换机均通过访问服务器来进行控制,该拓扑可以让 1~7 人共同操作。

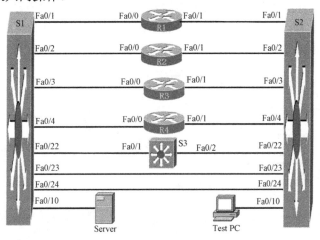

图 1-1 本书实验拓扑(以太网连接部分)

在图 1-1 拓扑中,4 台路由器均为 CISCO2911 路由器,IOS 采用 c2900-universalk9-mz.SPA.157-3.M.bin;3 台三层交换机为 Catalyst 3560 V2,IOS 采用 c3560-ipservicesk9-mz.150-2.SE11.bin 。所有路由器的 GigabitEthernet0/0 以太网端口与交换机 S1 进行连接;GigabitEthernet0/1 以太网接口则与交换机 S2 进行连接。交换机 S1 和 S2 之间通过 FastEthernet0/23 和 FastEthernet0/24 进行连接;交换机 S3 的 FastEthernet0/1 端口连接到 S1 的 FastEthernet0/22 上,FastEthernet0/2 端口连接到 S2 的 FastEthernet0/22 上。为了便于测试,

在图 1-1 中还连接了一台服务器和一台 PC。

4 台路由器之间通过串行链路进行连接，如图 1-2 拓扑所示。

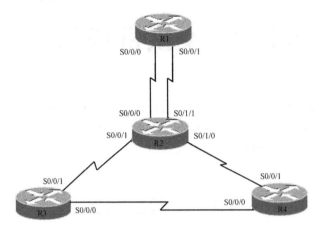

图 1-2　本书实验拓扑（广域网连接部分）

1.1.2　访问服务器

稍微复杂一点的实验就会用到多台路由器或者交换机，如果通过计算机的 COM 口和它们的 Console 口连接，由于一个 COM 口只能连接一台设备，就需要多台计算机或者经常拔插 Console 线，非常不方便。访问服务器可以解决这个问题，访问服务器和网络设备的连接方法如图 1-3 所示。访问服务器可以是一台插有 8 个（NM-8A 模块）或者 16 个（NM-16A 模块）异步口的路由器，从它引出多条连接线到各个路由器上（被控设备）的 Console 口。在使用时，用户首先 Telnet 到访问服务器，然后再从访问服务器访问各个路由器和交换机等被控设备，这样就能同时控制多台设备。

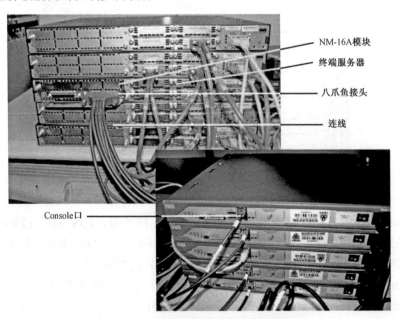

图 1-3　访问服务器和网络设备的连接方法

1.2 实验1：配置访问服务器

使用访问服务器（就是插有异步模块 NM-8A 或者 NM-16A 的路由器）可以避免在同时配置多台路由器时频繁拔插 Console 线，为了方便使用访问服务器，可以制作一个简单的菜单。

1. 实验目的

通过本实验可以掌握：
① 访问服务器的配置方法并制作一个简单的菜单。
② 访问服务器和交换机的使用方法。

2. 实验拓扑

访问服务器与各路由器和交换机连接实验拓扑如图 1-4 所示。

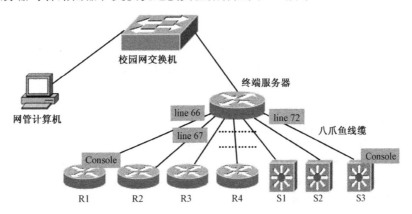

图 1-4 访问服务器与各路由器和交换机连接实验拓扑

3. 实验步骤

（1）访问服务器的基本配置

```
Router(config)#hostname Terminal-Server    //配置访问服务器的主机名
Terminal-Server(config)#enable secret CISCO
//配置进入特权模式的密码，防止他人修改访问服务器的配置
Terminal-Server(config)#no ip domain-lookup
//禁止路由器查找 DNS 服务器，防止输入错误命令时的长时间等待
Terminal-Server(config)#line vty 0 ?
    <1-15>   Last Line number
    <cr>
//查看该路由器支持多少 vty 虚拟终端，可以看到支持 0～988 个。路由器支持多少 vty 和路由器的 IOS 有关
Terminal-Server(config)#line vty 0 15
Terminal-Server(config-line)#login
```

Terminal-Server(config-line)#**password CISCO**
Terminal-Server(config-line)#**logging synchronous**
Terminal-Server(config-line)#**exec-timeout 0 0**
Terminal-Server(config-line)#**exit**
//以上配置 Telnet 该访问服务器时需要输入的密码，并且配置长时间不输入命令也不会自动 Logout 出来
Terminal-Server#**configure terminal**
Enter configuration commands, one per line. End with CNTL/Z.
Terminal-Server(config)#**interface fastEthernet 0/0**
Terminal-Server(config-if)#**ip address 10.3.24.31 255.255.255.0**
Terminal-Server(config-if)#**no shutdown**
Terminal-Server(config-if)#**exit**
//配置以太网端口的 IP 地址并打开端口
Terminal-Server(config)#**no ip routing**
//由于访问服务器不需要路由功能，所以关闭路由功能，这时访问服务器相当于一台计算机
Terminal-Server(config)#**ip default-gateway 10.3.24.254**
//配置网关，允许他人从别的网段 Telnet 该访问服务器

（2）配置线路，制作简易菜单

Terminal-Server#**show line**

	Tty	Line	Typ	Tx/Rx	A	Modem	Roty	AccO	AccI	Uses	Noise	Overruns	Int
*	0	0	CTY		-	-	-	-	-	4	0	0/0	-
	1	1	AUX	9600/9600	-	-	-	-	-	0	0	0/0	-
*	1/0	66	TTY	9600/9600	-	-	-	-	-	14	1	0/0	-
*	1/1	67	TTY	9600/9600	-	-	-	-	-	13	2790	0/0	-
*	1/2	68	TTY	9600/9600	-	-	-	-	-	12	7	4/16	-
	1/3	69	TTY	9600/9600	-	-	-	-	-	12	2	0/0	-
*	1/4	70	TTY	9600/9600	-	-	-	-	-	4	55	0/0	-
（省略部分输出）													
	1/14	80	TTY	9600/9600	-	-	-	-	-	0	0	0/0	-
	1/15	81	TTY	9600/9600	-	-	-	-	-	0	0	0/0	-
*	514	514	VTY		-	-	-	-	-	21	0	0/0	-
（省略部分输出）													
	520	520	VTY		-	-	-	-	-	6	0	0/0	-

以上是查看访问服务器上异步模块的各异步口所在的线路编号，TTY 表示的就是异步模块，该访问服务器模块有 16 个端口，线路编号为 66~81，这里实际上只用了 66~72。记住线路的编号，后面须要根据这些编号进行配置。

Terminal-Server#**configure terminal**
Terminal-Server(config)#**line 66 81**
Terminal-Server(config-line)#**transport input all**
//进入线路模式，线路允许所有进入，实际上只允许 Telnet 进入即可
Terminal-Server(config-line)#**no exec**
//不允许该 line 接受一个 exec 会话，即只能被反向 Telnet
Terminal-Server(config-line)#**exec-timeout 0 0**

//以上配置超时时间为 0
Terminal-Server(config-line)#**logging synchronous**
Terminal-Server(config-line)#**exit**
Terminal-Server(config)#**interface loopback0**
Terminal-Server(config-if)#**ip address 1.1.1.1 255.255.255.255**

创建一个环回口并配置 loopback0 端口的 IP 地址，loopback 端口是一个逻辑上的端口，路由器上可以任意创建几乎无穷多的 loopback 端口，该端口可以永远是 UP 的。loopback 端口经常用于测试等。

Terminal-Server(config)#**ip host R1 2066 1.1.1.1**
Terminal-Server(config)#**ip host R2 2067 1.1.1.1**
Terminal-Server(config)#**ip host R3 2068 1.1.1.1**
Terminal-Server(config)#**ip host R4 2069 1.1.1.1**
Terminal-Server(config)#**ip host S1 2070 1.1.1.1**
Terminal-Server(config)#**ip host S2 2071 1.1.1.1**
Terminal-Server(config)#**ip host S3 2072 1.1.1.1**

从访问服务器上控制各路由器是通过反向 Telnet 实现的，此时 Telnet 的端口号为线路编号加上 2000，例如，line 66，其端口号为 2066，如果要控制 line 66 线路上连接的路由器，可以采用"**telnet 1.1.1.1 2066**"命令。然而这样命令很长，为了方便，使用"**ip host**"命令定义一系列的主机名，这样可以直接输入"R1"控制 line 66 线路上连接的路由器了。

Terminal-Server(config)#**alias exec cr1 clear line 66**
Terminal-Server(config)#**alias exec cr2 clear line 67**
Terminal-Server(config)#**alias exec cr3 clear line 68**
Terminal-Server(config)#**alias exec cr4 clear line 69**
Terminal-Server(config)#**alias exec cs1 clear line 70**
Terminal-Server(config)#**alias exec cs2 clear line 71**
Terminal-Server(config)#**alias exec cs3 clear line 72**

//定义了一系列的命令别名，例如，"cr1" = "clear line 66"，"clear line"命令的作用是清除线路，有时候会出现无法连接到被控设备的情形，需要把线路清除一下

Terminal-Server(config)#**privilege exec level 0 clear line**
Terminal-Server(config)#**privilege exec level 0 clear**
//在用户模式下也能使用"clear line"和"clear"命令

Terminal-Server(config)#**banner motd @**
Enter TEXT message.   End with the character '@'.
```
        ******************************
        R1-------R1       cr1------clear line 66
        R2-------R2       cr2------clear line 67
        R3-------R3       cr3------clear line 68
        R4-------R4       cr4------clear line 69
        S1-------s1       cs1------clear line 70
        S2-------s2       cs2------clear line 71
        S3-------s3       cs3------clear line 72
        ******************************
```
@

以上命令完成了一个简单的菜单，提醒用户：要控制路由器 R1 可以使用"R1"命令（大小写不敏感）；要清除路由器 R1 所在的线路，可以使用"cr1"命令。这里是利用路由器的 banner motd 功能实现的，该功能使得用户 Telnet 到路由器后，就显示以上简易菜单。

Terminal-Server#copy   running-config   startup-config   //保存配置

4．实验调试

（1）测试能否从访问服务器上控制路由器和交换机

在计算机上配置网卡的 IP 地址为 10.3.24.60/255.255.255.0 网段上的 IP 地址，打开 DOS 命令行窗口。首先测试计算机和路由器的 IP 连通性，再进行 Telnet 远程登录。测试如下：

C:\Documents and Settings\longkey>**ping 10.3.24.31**
Pinging 10.3.24.31 with 32 bytes of data:
Reply from 10.3.24.31: bytes=32 time<1ms TTL=255
Reply from 10.3.24.31: bytes=32 time<1ms TTL=255
Reply from 10.3.24.31: bytes=32 time=1ms TTL=255
Reply from 10.3.24.31: bytes=32 time=18ms TTL=25

Ping statistics for 10.3.24.31:
　　Packets: Sent = 4, Received = 4, Lost = 0 (0%)
Approximate round trip times in milli-seconds:
Minimum = 0ms, Maximum = 18ms, Average = 4ms
//表明计算机能 ping 通访问服务器

C:\Documents and Settings\longkey>**telnet 10.3.24.31**
User Access Verification
Password:在此输入密码 **CISCO**
　　　　**
　　　　R1--------R1　　　cr1------clear line 66
　　　　R2--------R2　　　cr2------clear line 67
　　　　R3--------R3　　　cr3------clear line 68
　　　　R4--------R4　　　cr4------clear line 69
　　　　S1--------s1　　　cs1------clear line 70
　　　　S2--------s2　　　cs2------clear line 71
　　　　S3--------s3　　　cs3------clear line 72
　　　　**
//Telnet 到 10.3.24.31 后，出现简易菜单

Terminal-Server>**cr1**
[confirm]
 [OK]
Terminal-Server>   //先用"cr1"命令清除线路 66，该线路上连接了路由器 R1
Terminal-Server>**r1**
Trying R1 (1.1.1.1, 2066)... Open
　　　　**
　　　　R1--------R1　　　cr1------clear line 66
　　　　R2--------R2　　　cr2------clear line 67

```
          R3--------R3       cr3------clear line 68
          R4--------R4       cr4------clear line 69
          S1--------s1       cs1------clear line 70
          S2--------s2       cs2------clear line 71
          S3--------s3       cs3------clear line 72
          *******************************************
R1>
```

输入"r1"命令，如果出现"R1>"或者"Router>"等字符，表明可以控制路由器 R1 了；如果出现以下情况：

```
Terminal-Server>r1
Trying R1 (1.1.1.1, 2066)...
% Connection refused by remote host
```

在执行几次"cr1"命令后，重新执行"r1"命令。

（2）测试能否从访问服务器上控制各路由器和交换机

重复步骤（1），可以打开不同路由器或者交换机的控制窗口，这样就可以在一台计算机上同时配置不同的路由器和交换机了，如图 1-5 所示。当然，一台路由器只能被一台计算机所控制。

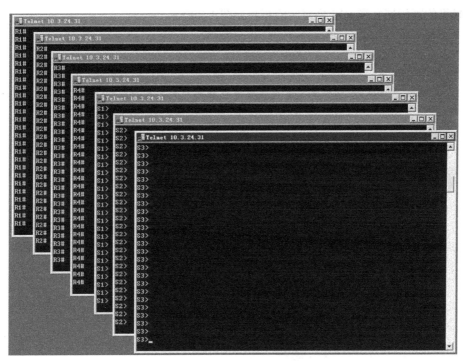

图 1-5 打开多个路由器或者交换机的控制窗口

提示

在实际应用中，如果须要配置多台设备，不建议使用 Windows 自带的 Telnet 程序，可以选用 SecureCRT 等专业终端软件，这些软件的功能完善，更方便使用。使用 SecureCRT 软件打开多个路由器或者交换机的控制窗口，如图 1-6 所示。

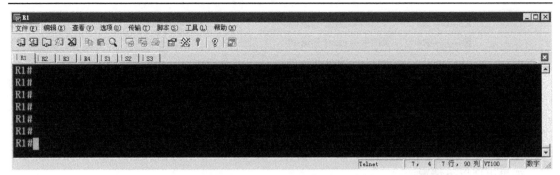

图 1-6　使用 SecureCRT 软件打开多个路由器或者交换机的控制窗口

1.3　实验 2：交换机的密码恢复

1. 实验目的

通过本实验可以掌握交换机的密码恢复步骤。

2. 实验拓扑

交换机的密码恢复和 IOS 恢复实验拓扑如图 1-7 所示。

图 1-7　交换机的密码恢复和 IOS 恢复实验拓扑

3. 实验步骤

　　Cisco 交换机的密码恢复步骤和路由器的密码恢复方法差别较大，并且不同型号的交换机恢复方法也有所差异。和路由器一样，在恢复交换机密码的过程中操作者也要在交换机的现场。以下是 Catalyst 3560 V2（Catalyst 2950 也类似）交换机的密码恢复步骤。

　　① 拔掉交换机电源，按住交换机前面板的 Mode 键不放，接上电源，此时，你会看到如下提示。

```
Base ethernet MAC Address: 00:23:ac:7d:6c:80
Xmodem file system is available.
The password-recovery mechanism is enabled.

The system has been interrupted prior to initializing the
flash filesystem.   The following commands will initialize
the flash filesystem, and finish loading the operating
system software:
    flash_init
```

boot

② 输入 **flash_init** 命令。

switch: **flash_init**
Initializing Flash...
flashfs[0]: 456 files, 6 directories
flashfs[0]: 0 orphaned files, 0 orphaned directories
flashfs[0]: Total bytes: 32514048
flashfs[0]: Bytes used: 17016320
flashfs[0]: Bytes available: 15497728
flashfs[0]: flashfs fsck took 14 seconds.
...done Initializing Flash.

③ 修改配置文件名。

switch: **dir flash:**
Directory of flash:/
2   -rwx   1639   <date>   config.text
4   -rwx   7960810   <date>   c3560-advipservicesk9-mz.122-46.SE.bin
15497728 bytes available (17016320 bytes used)
//config.text 是交换机的启动配置文件，和路由器的 startup-config 类似

switch: **rename flash:config.text flash:config.old**
//修改启动配置文件名，这样交换机在启动时就读不到 config.text 了，交换机启动后将没有任何配置，从而没有了密码

④ 输入 **boot** 命令重启系统，这次就不要再按住 Mode 键了。启动需要等待几分钟时间。

⑤ 当出现如下提示时，输入 n。

Would you like to enter the initial configuration dialog? [yes/no]:n

⑥ 用 **enable** 命令进入 enable 状态，并将文件 config.old 改回 config.text。

Switch#**rename flash:config.old flash:config.text**
Destination filename [config.text]?回车

⑦ 将原配置装入内存。

Switch#**copy   flash:config.text   running-config**
Destination filename [running-config]?回车

⑧ 修改密码。

S1#**configure terminal**
S1(config)#**enable secret cisco**
S1(config)#**exit**

⑨ 将配置重新写入 nvram。

S1#**copy running-config startup-config**
Destination filename [startup-config]?回车

1.4 实验 3：交换机的 IOS 恢复

1. 实验目的

通过本实验可以掌握交换机的 IOS 恢复步骤。

2. 实验拓扑

交换机的 IOS 恢复实验拓扑如图 1-7 所示。为了节约时间，本实验使用了一个较小的 IOS c3560-advipservicesk9-mz.122-46.SE.bin 进行恢复。

3. 实验步骤

如果交换机已经正常开机，IOS 不小心被破坏，则 IOS 可以从 TFTP 服务器上恢复（使用 **copy tftp flash** 命令，本书不在此介绍该方法）。然而，如果交换机无法正常开机，交换机 IOS 不能从 TFTP 服务器恢复，而要使用 XModem 方式，该方式是通过 Console 口从计算机下载 IOS，因此速度很慢。步骤如下：

① 把计算机的串口和交换机的 Console 口连接好，用超级终端软件连接上交换机，默认时 Console 的通信速率是 9 600 bps。

② 交换机开机后（因为 IOS 有故障，所以无法正常开机），执行以下命令。

```
Interrupt within 5 seconds to abort boot process.
Boot process failed...

The system is unable to boot automatically.    The BOOT
environment variable needs to be set to a bootable
image.
switch: flash_init
```

由于默认时 Console 的波特率（通信速率）是 9 600 bps，如果用该速率来传送 IOS 会很费时，所以需要把 Console 的波特率改为 115 200 bps，如下：

```
switch: set BAUD 115200
```

这将造成超级终端软件和交换机的连接断开，请用新的速率 115 200 bps，重新把超级终端软件和交换机进行连接。

③ 输入拷贝指令。

```
switch:copy -b 4096 xmodem: flash:c3560-advipservicesk9-mz.122-46.SE.bin
```

该命令的含义是通过 Xmodem 方式拷贝文件，保存在 Flash 中，文件名为 **c3560-advipservicesk9-mz.122-46.SE.bin**，4096 是缓冲区的大小。出现如下提示：

```
Begin the Xmodem or Xmodem-1K transfer now...
CCCC
```

在超级终端窗口中，选择【发送】→【发送文件】菜单，打开图 1-8 窗口，选择 IOS 文件，协议为"Xmodem"。单击"发送"按钮开始发送文件。由于速度很慢，通常需要几个小时，请耐心等待，通信速率为 115 200 bps，如图 1-9 所示。

第 1 章 交换机基本配置

图 1-8 选择 IOS 文件

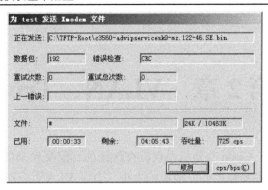

图 1-9 正在发送 IOS

④ 传送完毕后执行以下命令。

需要把 Console 的通信速率恢复为 9 600 bps，如下：

switch: **unset BAUD**

同样将造成超级终端软件和交换机的连接断开，请用新的速率 9 600 bps 将超级终端软件和交换机进行连接。

switch:**boot**

启动系统。

 提示

交换机和路由不同，其 IOS 如果被删除或者损坏，则只能通过 Xmodem 的方式恢复，所需时间很长，因此要尽量避免 IOS 的毁坏。如果是升级 IOS，请在交换机正常开机后，使用 **copy tftp flash** 命令来升级，这时是通过以太网来传输 IOS 的，速度会快得多。

1.5 本章小结

本章介绍了如何通过把计算机上的串口和交换机的 Console 进行连接来配置交换机，还介绍了如何配置访问服务器，方便我们同时配置多台路由器或者交换机，最后还介绍了交换机的密码和 IOS 恢复步骤。本章给出了一直贯穿本书的网络拓扑。

第 2 章　VLAN、Trunk、VTP 与链路聚集

Cisco 交换机不仅具有传统的二层交换功能，它还具有 VLAN 等功能。VLAN 技术可以很容易地控制广播域的大小。有了 VLAN，交换机之间的级联链路就需要 Trunk 技术来保证该链路可以同时传输多个 VLAN 的数据。管理员可以手动配置交换机之间链路上的 Trunk，也可以让交换机自动协商，协商的协议称为 DTP。交换机之间的级联链路带宽如果不够，可以把多条链路捆绑起来形成逻辑链路。在稍微大型一点的网络中，会有多个交换机，同时也会有多个 VLAN，如果在每个交换机上分别把 VLAN 创建一遍，工作量很大，VTP 协议可以解决这个问题，管理员一次性创建或修改 VLAN，VTP 会把 VLAN 信息自动同步到所有的交换机上。私有 VLAN 采用两层 VLAN 隔离技术，只有上层 VLAN 全局可见，下层 VLAN 相互隔离。私有 VLAN 通常用于企业内部网，用来防止连接到某些端口或端口组的网络设备之间的相互通信，但却允许与默认网关进行通信，尽管各设备处于不同的私有 VLAN 中，它们可以使用相同的 IP 子网。

2.1　VLAN、Trunk、VTP 与链路聚集概述

2.1.1　交换机工作原理

传统的交换机二层设备可以隔离冲突域，每个端口是单独的冲突域。交换机是基于收到的数据帧中的源 MAC 地址和目的 MAC 地址来进行工作的。交换机的作用主要有两个：一个是维护 CAM（Context Address Memory）表，该表是 MAC 地址和交换机端口的映射表；另一个是根据 CAM 来进行数据帧的转发。

交换机有如下 5 种基本操作。

① 获取（学习）：当交换机从某个端口收到数据帧时，交换机会读取帧的源 MAC 地址，并在 MAC 表中填入 MAC 地址及其对应的端口。

② 过期：通过获取过程获取的 MAC 表条目具有时间戳，此时间戳用于从 MAC 表中删除旧条目。当某个条目在 MAC 表中创建之后，就会使用其时间戳作为起始值开始递减计数。当计数为 0 时，条目被删除。交换机当从相同端口接收同一源 MAC 的帧时，将会刷新表中的该条目。

③ 泛洪：如果目的 MAC 地址不在 MAC 表中，交换机就不知道从哪一个端口发送帧，此时它会将帧发送到除接收端口以外的所有其他端口，这个过程称为泛洪。泛洪还用于发送目的地址为广播或者组播 MAC 地址的帧。

④ 选择性转发：选择性转发是检查帧目的 MAC 地址后，将帧从适当的端口转发出去。

当计算机发送帧到交换机时,如果交换机知道该节点的 MAC 地址,交换机会将此地址与 MAC 表中的条目比对,然后将帧转发到相应的端口。此时交换机不是将帧泛洪到所有端口,而是通过其指定端口发送到目的计算机。

⑤ **过滤**:在某些情况下,帧不会被转发,此过程称为帧过滤。前面已经描述了过滤的使用——交换机不会将帧转发到接收帧的端口。另外,交换机还会丢弃损坏的帧。如果帧没有通过 CRC 检查,就会被丢弃。对帧进行过滤的另一个原因是安全,交换机具有安全设置,用于阻挡发往或来自选定 MAC 地址或特定端口的帧。

2.1.2 VLAN 简介

交换机能够隔离冲突域,然而不能隔离广播域,通过多个交换机连接在一起的所有计算机都在一个广播域中,任何一台计算机发送广播包,其他计算机都会收到,这样大大降低了带宽的利用率,同时也使得计算机 CPU 忙于处理不需要的广播包。VLAN(Virtual LAN)可以隔离广播域。VLAN 工作在 OSI 的第二层,VLAN 是交换机端口的逻辑组合,可以把在同一交换上的端口组合成一个 VLAN,也可以把在不同交换上的端口组合成一个 VLAN,一个 VLAN 就是一个广播域,VLAN 之间的通信如果不通过第三层的路由器(或者三层交换机)是无法实现的。

1. VLAN 主要的优点

① **广播风暴防范**:将网络划分为多个 VLAN 可减少参与广播风暴的设备数量。每一个 VLAN 是一个广播域,这样每个广播域中的计算机数量就大为减少。通常每个 VLAN 是一个独立的 IP 子网。

② **安全**:含有敏感数据的用户组可与网络的其余部分隔离,从而降低泄露机密信息的可能性。

③ **性能提高**:将第二层平面网络划分为多个逻辑工作组(广播域)可以减少网络上不必要的流量并提高性能。

④ **提高管理效率**:由于 VLAN 是逻辑上的组合,管理员可以很容易地通过修改配置重新划分 VLAN,而不须要改变物理拓扑,大大提高了管理的效率。

2. 划分 VLAN 的方法

① 基于端口的 VLAN(静态 VLAN):管理员可以用手动方式把交换机某一端口指定为某一 VLAN 的成员,绝大数工程中采用的 VLAN 划分方法是该种方法。

② 基于 MAC 地址的 VLAN(动态 VLAN):通过一种称为 VLAN 成员资格策略服务器(VMPS)的特殊服务器完成。使用 VMPS 可以根据连接到交换机端口的设备的源 MAC 地址,动态地将端口分配给 VLAN。即使将计算机从一台交换机的端口移到另一台交换机的端口,第二台交换机也会将该主机的端口动态地分配给相同的 VLAN。除了基于 MAC 地址的 VLAN,还有基于 IP 地址的 VLAN。这些方式不常见,配置起来还相当麻烦,本书不进行介绍。

2.1.3 Trunk 简介

当一个 VLAN 跨过不同的交换机时，在同一 VLAN 上、但是接在不同的交换机上的计算机需要通信时将如何实现呢？可以在交换机之间为每一个 VLAN 都增加连线，然而这样的方法在有多个 VLAN 时会占用太多的以太网端口。可以采用 Trunk 技术实现跨交换机的 VLAN 内通信，Trunk 技术使得在一条物理线路上可以传送多个 VLAN 的数据，交换机从属于某一 VLAN（如 VLAN3）的端口接收到数据，在 Trunk 链路上进行传输前，会加上一个标记，表明该数据是 VLAN3 的；到了对方交换机，交换机会把该标记去掉，只发送到属于 VLAN3 的端口上。

有两种常见的 Trunk 帧标记技术：ISL 和 IEEE 802.1q。ISL 技术在原有的帧上重新加了一个帧头，并重新生成了帧校验序列（FCS），ISL 是思科特有的技术，因此不能在 Cisco 交换机和非 Cisco 交换机之间使用。图 2-1 是 ISL 的封装。

① **DA**：48 位的目的 MAC 地址，为 01-00-0c-00-00。
② **TYPE**：类型，4 位，表明被封装的帧的类型，将来还可以用来指示不同的封装类型。0000——以太网；0001——令牌环；0010——fddi；0011：atm。
③ **USER**：用户定义位，4 位，TYPE 的一种扩展，也用来表明域的类型。
④ **SA**：源地址，48 位的 MAC 地址。
⑤ **LEN**：长度，16 位，用来代表数据包的长度，长度不包含 CRC 位。
⑥ **SNAP/LLC**：24 位的一个常数值，为 AAAA03。
⑦ **HSA**：24 位，生产厂商的 ID 部分 SA 的前 3 字节。
⑧ **VLAN ID**：15 位，标示帧的所属 VLAN ID。
⑨ **BPDU/CDP 指示符**：1 位，BPDU 置位，用来表示网络拓扑的改变情况，STP 协议用其协商。
⑩ **INDEX（索引）**：16 位，数据离开交换机端口的端口索引，只用于诊断的目的。
⑪ **FDDI 和令牌环的保留域（RES）**：16 位，关于 FDDI 经过 ISL 封装后的帧的变化情况。
⑫ **FCS**：帧校验序列，32 位 CRC 值，对 DA 到整个封装完成后的帧进行计算，接收方交换机校验该 CRC，并可以丢弃那些 CRC 无效的数据包。

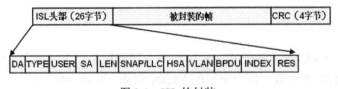

图 2-1 ISL 的封装

IEEE 802.1q 技术在原有帧的源 MAC 地址字段后插入 4 字节的标记字段，同时用新的 FCS 字段替代了原有的 FCS 字段，IEEE 802.1q 的封装如图 2-2 所示。该技术是国际标准，得到所有厂家的支持。

① **TPID**：16 位，协议标示符，包含一个 0x8100 的固定值，它表明该帧带有 IEEE 802.1q/IEEE 802.1p 的标记信息。
② **PRI**：3 位，IEEE 802.1q 优先级。

③ CFI：1 位的规范形式表示符，以太网交换机为 0，令牌环为 1。
④ VLAN ID：12 位，标示帧的所属 VLAN ID。

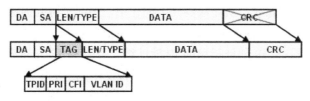

图 2-2　IEEE 802.1q 的封装

Trunk 链路上把数据重新封装，将直接导致帧头变大，从而影响效率，可以在 Trunk 链路上指定一个 Native VLAN（本征 VLAN，默认是 VLAN1），这样来自 Native VLAN 的数据帧通过 Trunk 链路时将不重新封装，以原有的帧传输。显然 Trunk 链路的两端指定的 Native VLAN 要一致，否则将导致数据帧从一个 VLAN 传播到另一个 VLAN 上。

2.1.4　DTP 简介

管理员可以用手动的方法指定交换机之间的链路是否形成 Trunk，也可以让 Cisco 交换机自动协商，这个自动协商的协议称为 DTP（Dynamic Trunk Protocol），DTP 还可以协商 Trunk 链路的封装类型。配置了 DTP 的交换机会发送 DTP 协商包，或者对对方发送来的 DTP 包进行响应，双方交换机最终一致同意它们之间的链路是否形成 Trunk，以及采用什么样的 Trunk 封装方式。双方的交换机的 DTP 配置只有在一些情况下才能成功协商形成 Trunk，表 2-1 是链路两端是否会形成 Trunk 的 DTP 总结。

表 2-1　DTP 总结

	negotiate	desirable	auto	nonegotiate
negotiate	√	√	√	√
desirable	√	√	√	×
auto	√	√	×	×
nonegotiate	√	×	×	√

在以上模式中，negotiate 模式已经把端口强制置成 Trunk 模式，并会主动请求对方启用 Trunk，也会响应对方的请求；desirable 模式期望把端口置成 Trunk 模式，会主动请求对方启用 Trunk，也会响应对方的请求，只要对方响应请求则会成功协商成 Trunk；auto 模式不会主动请求对方启用 Trunk，但是会响应对方的请求，如果对方主动请求则会成功协商成 Trunk；nonegotiate 模式已经把端口强制置成 Trunk 模式，但是不主动请求对方启用 Trunk，也不会响应对方的请求，除非对方也已经把端口强制置成 Trunk 模式，否则无法形成 Trunk。

2.1.5　EtherChannel 简介

EtherChannel（以太通道）是交换机之间的多链路捆绑技术。它的基本原理是：将两台设备间多条快速以太或千兆位以太物理链路捆绑在一起组成一条逻辑链路，从而达到带宽倍增的目的。除了增加带宽，EtherChannel 还可以在多条链路上均衡分配流量，起到负载分担的作用；当一条或多条链路发生故障时，只要还有链路正常，流量将转移到其他的链路上，整个过程在很短的时间里完成，从而起到冗余的作用，增强了网络的稳定性和安全性。

在 EtherChannel 中，负载在各个链路上的分布可以根据源 IP 地址、目的 IP 地址、源 MAC 地址、目的 MAC 地址、源 IP 地址和目的 IP 地址组合及源 MAC 地址和目的 MAC 地址组合等来进行分布。EtherChannel 分发流量的方法是对报文中选定的字段（源 MAC 地址和目的 MAC 地址、源 IP 地址和目的 IP 地址等）进行 Hash 运算，生成 0~7 的数值，并根据 Hash 结果将报文发送到物理链路上。EtherChannel 不一定能把流量平均分配到各物理链路上。例子 1：有 3 条链路，则即使流量是随机流量，各物理链路的流量比例为 3∶3∶2；例子 2：有 2 条链路，采用基于目的 IP 的负载均衡，则到达同一服务器的全部流量只会从一固定链路上发出，而另一链路上没有流量。

EtherChannel（以太通道）可以把 Access 或者 Trunk 端口进行捆绑，也可以捆绑三层端口，但是要求同一 EtherChannel 中的物理端口有同一特性。同一 EtherChannel 中的物理端口要么都是 Access 端口，要么都是 Trunk 端口，要么都是三层端口。一个 EtherChannel 最多可以捆绑 16 个端口，其中最多 8 个端口是活动的。在配置 EtherChannel 时，同一组中的全部端口配置应该相同，即不能把 Trunk 端口和 Access 端口进行混合捆绑。

两台交换机之间是否形成 EtherChannel 也可以用协议自动协商。目前有两个协商协议：PAGP 和 LACP，前者是 Cisco 专有的协议，而 LACP 是公共标准。在这两个协议中，端口有不同的模式，不同模式组合会有不同的协商结果。表 2-2 是 PAGP 协商的规律总结，表 2-3 是 LACP 协商的规律总结。两个表中的"ON"表示管理员手动配置了 EtherChannel。

表 2-2 PAGP 协商的规律总结

	ON	desirable	auto
ON	√	×	×
desirable	×	√	√
auto	×	√	×

表 2-3 LACP 协商的规律总结

	ON	active	passive
ON	√	×	×
active	×	√	√
passive	×	√	×

2.1.6 VTP

1. 为什么需要 VTP

假设网络中有 M 个交换机，网络中划分 N 个 VLAN。则为保证网络正常工作，需要在每个交换机上都创建 N 个 VLAN，共 $M×N$ 个 VLAN，随着 M 和 N 的增大，这将是一项枯燥而繁重的任务。VTP（VLAN Trunk Protocol）可以帮助管理员减少这些枯燥繁重的工作。管理员在网络中设置一个或者多个 VTP Server，然后在 Server 上创建、修改 VLAN，VTP 协议会将这些修改通告其他交换机上，这些交换机自动更新 VLAN 的信息（就是 VLAN ID 和 VLAN 的名字）。VTP 使得 VLAN 的管理自动化了。

2. VTP 域与 VTP 角色

所谓的 VTP 域（VTP Domain）由需要共享相同 VLAN 信息的交换机组成，只有在同一 VTP 域（即 VTP 域的名字相同）的交换机才能同步 VLAN 信息。根据交换机在 VTP 域中的作用不同，VTP 可以分为 3 种模式，表 2-4 是 VTP 模式比较。

① 服务器模式（Server）：在 VTP 服务器上能创建、修改、删除 VLAN，同时这些信息会通告给域中的其他交换机；VTP 服务器收到其他交换机的 VTP 通告后会更改自己的 VLAN 信息并进行转发。VTP 服务器会把 VLAN 信息保存在 NVRAM 中，即 flash:vlan.dat 文件，即使重新启动交换机，这些 VLAN 还会存在。默认情况下，交换机是服务器模式。每个 VTP 域必须至少有一台服务器，当然也可以有多台。

② 客户机模式（Client）：VTP 客户机上不允许创建、修改、删除 VLAN，但它会监听来自其他交换机的 VTP 通告并更改自己的 VLAN 信息，接收到的 VTP 信息也会在 Trunk 链路上向其他交换机转发，因此这种交换机还能充当 VTP 中继；VTP Client 把 VLAN 信息保存在 RAM 中，重新启动后这些信息会丢失。

③ 透明模式（Transparent）：这种模式的交换机不完全参与 VTP。可以在这种模式的交换机上创建、修改、删除 VLAN，但是这些 VLAN 信息并不会通告给其他交换机，它也不接收其他交换机的 VTP 通告而更新自己的 VLAN 信息。然而需要注意的是，它会通过 Trunk 链路转发接收到的 VTP 通告，从而充当了 VTP 中继的角色，因此完全可以把该交换机看成透明的。VTP Transparent 仅会把本交换机上 VLAN 信息保存在 NVRAM 中。

表 2-4 VTP 模式比较

模式	能创建、修改、删除 VLAN	能转发 VTP 信息	会根据收到 VTP 包更改 VLAN 信息	会保存 VLAN 信息	会影响其他交换机上的 VLAN
Server（服务器）	是	是	是	是	是
Client（客户）	否	是	是	否（见提示）	是
Transparent（透明）	是	是	否	是	否

> **提示**
> 根据 Cisco 的资料，VTP Client 应该不会在 flash:vlan.dat 文件存放 VLAN 信息，然而多次的实验结果和这一说法并不一致。

3. VTP 通告

VLAN 信息的同步是通过 VTP 通告来实现的，VTP 通告只能在 Trunk 链路上传输（因此交换机之间的链路必须成功配置了 Trunk）。VTP 通告是以组播帧的方式发送的，VTP 通告中有一个字段称为修订号（Revision），代表 VTP 帧的修订级别，它是一个 32 位的数字。交换机的默认配置号为零。在每次添加或删除 VLAN 时，修订号都会递增。修订号用于确定从另一台交换机收到的 VLAN 信息是否比储存在本交换机上的信息更新。如果收到的 VTP 通告修订号更大，则本交换机将根据此通告更新自身的 VLAN 信息。如果交换机收到修订号更小的通告，会用自己的 VLAN 信息反向覆盖。需要注意的是，大修订号的通告会覆盖小修订号的通告，而不管自己或者对方是 Server 还是 Client。

VTP 通告有总结通告、子集通告和请求通告几种类型。总结通告包含 VTP 域名、当前修订号和其他 VTP 配置详细信息。在以下情况下会发送总结通告：VTP 服务器或客户端每 5 分钟发送一次总结通告给邻居交换机，通告当前 VTP 修订号；执行配置操作后也会立即发送总结通告。总结通告的格式如图 2-3 所示。

① **版本：**1 字节，Version 1/ Version 2。

② **编码：**1 字节，表示 VTP 的消息类型。

③ **后续通告数：**1 字节，随后有多少个子集通告消息（0～255，0 为没有）。

④ **管理域名长度：**1 字节，表示 vtp 域名的长度。

⑤ **管理域名：**32 字节，表示域名。

⑥ **配置修订号：**4 字节。

⑦ **更新者标识：**4 字节，最近对修订号进行增加的交换机的 IP 地址。

⑧ **更新时间戳：**12 字节，最近的更改时间和日期。

⑨ **md5 摘要：**16 字节，包含了消息的哈希值。

0　　7	8　　15	16　　23	24　　31
版本	编码	后续通告数	管理域名长度
管理域名（32字节）			
配置修订号（4字节）			
更新者标识（4字节）			
更新时间戳（12字节）			
md5摘要（16字节）			

图 2-3　VTP 总结通告的格式

子集通告包含 VLAN 信息。触发子集通告的更改包括：创建或删除 VLAN、暂停或激活 VLAN、更改 VLAN 名称和更改 VLAN 的 MTU。可能需要多个子集通告才能完全更新 VLAN 信息。子集通告列出了每个 VLAN 的信息，包括默认 VLAN。子集通告的格式如图 2-4 所示。

① **编码：**0x02 表明是子集通告。

② **系列号：**代表数据包在汇总通告后的一系列的子集通告中的序列号。

0　　7	8　　15	16　　23	24　　31
版本	编码	系列号	管理域名长度
管理域名（32字节）			
配置修订号（4字节）			
VLAN信息字段1			
……			
VLAN信息字段N			

图 2-4　VTP 子集通告的格式

VTP VLAN 信息字段格式如图 2-5。

① **VLAN 的状态：**活动或者挂起。

② **VLAN 的类型**：以太网、令牌环和 FDDI 或者其他类型。
③ **MTU 的大小**：本 VLAN 所支持的最大 MTU。

0	7	8	15	16	23	24	31
信息长度		状态		VLAN类型		VLAN名字长度	
VLAN ID				MTU大小			
VLAN名称							

图 2-5　VTP VLAN 信息字段格式

当向相同 VTP 域中的 VTP 服务器发送请求通告时，VTP 服务器的响应方式是：先发送总结通告，接着送出子集通告。当发生以下情况时，将发送请求通告——VTP 域名变动，交换机收到的总结通告包含比自身更大的修订号，子集通告消息由于某些原因丢失，交换机被重置。VTP 请求通告格式如图 2-6 所示。

① **编码**：0x03 表明是请求通告。
② **Rsvd**：保留。
③ **起始值**：要请求的第 $N+1$ 个子集通告。

0	7	8	15	16	23	24	31
版本		编码		Rsvd		管理域名长度	
管理域名							
起始值							

图 2-6　VTP 请求通告格式

4. VTP 修剪

VTP 没有进行修剪时的情况如图 2-7 所示。当 Switch1 的 VLAN 1 上计算机 A 发送广播包时，广播包将在所有交换机的 Trunk 链路上传输，然而图中 Switch3、Switch5 和 Switch6 根本没有该 VLAN 的计算机，这样浪费了 Trunk 链路上的带宽。可以在交换机上启用 VTP 修剪功能，VTP 进行修剪时的情况如图 2-8 所示。Switch4 和 Switch5、Switch2 和 Switch3 交换机上的 Trunk 链路上不再有 VLAN1 的广播包了。VTP 修剪功能会自动计算哪些链路应该修剪哪些 VLAN 的数据包，管理员只须要启用该功能就可以了。VTP 修剪功能将大大提高带宽的利用率。

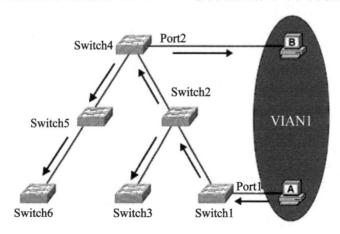

图 2-7　VTP 没有进行修剪时的情况

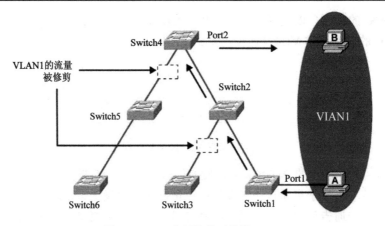

图 2-8　VTP 进行修剪时的情况

2.1.7　私有 VLAN

1. 为什么需要私有 VLAN

随着网络的迅速发展，用户对网络的安全性提出了更高的要求，传统的解决方法是给每个客户群分配一个 VLAN 和相关的 IP 子网，通过使用 VLAN，每个客户群被从第二层隔离开，可以防止任何恶意的行为和 Ethernet 的信息探听。然而，这种每个客户群单一 VLAN 和 IP 子网的模型造成了可扩展方面的局限性。例如，在图 2-9 中，要求 PC1 和 PC2 都能和 Server 进行通信，然而 PC1 和 PC2 之间不能通信。对于这个简单的需求，可以通过在交换机的 Fa0/1 和 Fa0/2 端口执行"**switchport protected**"命令来实现，其效果如图 2-9 所示。然而如果有更复杂的需求，则需要私有 VLAN 技术来实现。

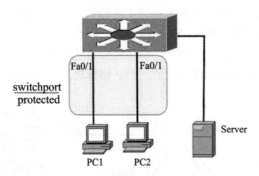

图 2-9　switchport protected 效果

2. 私有 VLAN

在私有 LAN 的概念中，交换机端口有 3 种类型：Isolated Port（隔离端口）、Community Port（团体端口）和 Promiscuous Port（混杂端口），它们分别对应不同的 VLAN 类型：Isolated Port 属于 Isolated PVLAN，Community Port 属于 Community PVLAN，而代表一个 Private VLAN 整体的是 Primary VLAN，前面两类 VLAN（Isolated PVLAN、Community PVLAN）也称为

辅助私有 VLAN，需要和 Primary VLAN 绑定在一起，Promiscuous Port 属于 Primary PVLAN。

在 Isolated PVLAN 中，Isolated Port 只能和 Promiscuous Port 通信，彼此不能交换流量；在 Community PVLAN 中，Community Port 不仅可以和 Promiscuous Port 通信，而且彼此也可以交换流量。Promiscuous Port 与路由器或三层交换机端口相连，可以与 Isolated Port 和 Community Port 通信。

PVLAN 的应用对于保证接入网络的数据通信的安全性是非常有效的，用户只需与自己的默认网关连接，一个 PVLAN 不需要多个 VLAN 和 IP 子网就能提供具备二层数据通信安全性的连接，所有的用户都接入 PVLAN，从而实现了所有用户与默认网关的连接，而与 PVLAN 内的其他用户没有任何访问。PVLAN 功能可以保证同一个 VLAN 中的各个端口相互之间不能通信，但可以穿过 Trunk 端口。这样即使同一 VLAN 中的用户相互之间也不会受到广播的影响。

2.2 实验 1：交换机基本配置

1. 实验目的

通过本实验可以掌握交换机的基本配置方法。

2. 实验拓扑

交换机基本配置实验拓扑如图 2-10 所示。

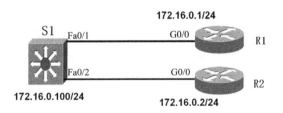

图 2-10　交换机基本配置实验拓扑

3. 实验步骤

（1）配置主机名

```
Switch>enable
Switch#configure terminal
Switch(config)#hostname S1

R1(config)#interface gigabitEthernet0/0
R1(config-if)#no shutdown
R1(config-if)#ip address 172.16.0.1 255.255.255.0
R2(config)#interface gigabitEthernet0/0
R2(config-if)#no shutdown
```

R2(config-if)#**ip address 172.16.0.2 255.255.255.0**
//R1 和 R2 的配置是为了测试

(2) 配置基本安全措施

S1(config)#**enable secret cisco**
S1(config)#**service password-encryption**
S1(config)#**line vty 0 15**
S1(config-line)#**password cisco**
S1(config-line)#**login**
S1(config)#**line console 0**
S1(config-line)#**password cisco**
S1(config-line)#**login**

(3) 完成端口基本配置

默认时,交换机的以太网端口是开启的;对于交换机的以太网口,可以配置双工模式及速率等。

S1(config)#**interface fastEthernet0/1**
S1(config-if)#**duplex full**
//duplex duplex { full | half | auto } 用来配置端口的双工模式,full——全双工,half——半双工,auto——自动检测双工的模式
S1(config-if)#**speed 100**
//speed { 10 | 100 | 1000 | auto }命令用来配置交换机的端口速度,10—10 Mbps,100—100 Mbps,1 000—1 000 Mbps,auto—自动检测端口速度
S1(config-if)#**mdix auto**
// 早期交换机端口在连接不同设备时需要使用不同电缆(例如,交换机对交换机需要使用交叉电缆,而交换机对路由器需要使用直通电缆)。使用"**mdix auto**"命令可以启用介质检测功能,这样端口上使用什么电缆都无所谓了
S1(config-if)#**description Connected to R1**
//在端口上配置描述,相当于 C 语言中的注释,描述并不影响端口的功能

(4) 配置管理地址

交换机也允许被 Telnet,这时需要在交换机上配置一个 IP 地址,这个地址是在 VLAN 端口上配置的(该地址称为管理地址),配置如下:

S1(config)#**interface vlan 1**
S1(config-if)#**ip address 172.16.0.100 255.255.255.0**
S1(config-if)#**no shutdown**
S1(config)#**ip default-gateway 172.16.0.254**

在 VLAN1 端口上配置了管理地址,接在 VLAN1 上的计算机可以直接 Telnet 该地址。为了使其他网段的计算机也可以 Telnet 交换机,在交换机上配置了默认网关。

(5) 配置 SSH

由于 Telnet 协议使用明文传输数据,采用 Telnet 管理交换机,可能会有安全问题。可以采用 SSH 来提高安全性。配置如下:

```
S1(config)#ip domain-name ccnp.cisco.com
//配置域名，生成密钥时需要有域名
S1(config)#crypto key generate rsa
% You already have RSA keys defined named S1.cisco.ccnp.com.
% Do you really want to replace them? [yes/no]:yes      //是否覆盖已经存在的密钥
//SSH 的关键字名就是 "hostname + . +ip domain-name"
Choose the size of the key modulus in the range of 360 to 2048 for your
    General Purpose Keys. Choosing a key modulus greater than 512 may take
    a few minutes.
How many bits in the modulus [512]:      //回答密钥的长度为 360～2 048 bit，密钥越长越安全、开销越大
% Generating 1024 bit RSA keys, keys will be non-exportable...
[OK] (elapsed time was 3 seconds)
//密钥的长度，长度可以为 360～2 048 bit，密钥越长安全性越高，然而性能越差。这里采用默认值
S1(config)# ip ssh version 2
//使用 SSH 的版本 2
S1(config)#line vty 0 15
S1(config-line)#login local
//SSH 需要用户名和密码，配置用户和密码放在本地，即在交换机的配置文件中
S1(config)#username test secret cisco
//配置用户名和密码
```

(6) 管理端口的错误条件

交换机的端口可能会遭遇各种各样的错误，如环路等，默认时交换机会关闭端口。管理员需要在故障原因排除后执行 "**shutdown**" 和 "**no shutdown**" 命令把端口重新打开。可以配置端口自动关闭的条件和端口在故障后的自动恢复。

```
S1(config)#errdisable detect cause ?
  all                  Enable error detection on all cases
  arp-inspection       Enable error detection for arp inspection
  dhcp-rate-limit      Enable error detection on dhcp-rate-limit
  dtp-flap             Enable error detection on dtp-flapping
  gbic-invalid         Enable error detection on gbic-invalid
  l2ptguard            Enable error detection on l2protocol-tunnel
  link-flap            Enable error detection on linkstate-flapping
  loopback             Enable error detection on loopback
  pagp-flap            Enable error detection on pagp-flapping
  sfp-config-mismatch  Enable error detection on SFP config mismatch
//配置什么情况下会自动关闭端口，默认时是 ALL

S1(config)#errdisable recovery cause bpduguard
//配置由于 bpduguard 引起的端口故障会自动恢复端口，默认时不会自动恢复端口，使用 "errdisable
recovery cause ?" 命令可以查看各种会引起端口关闭的原因
S1(config)#errdisable recovery cause all
//配置所有原因引起的端口故障都会自动恢复端口
```

```
S1(config)#errdisable recovery interval 30
```
//配置端口错误关闭后间隔多长时间会进行自动恢复。当然,需要在上一步骤配置什么情况会自动恢复

4. 实验调试

(1) show ip interface brief

```
S1#show ip interface brief
Interface              IP-Address      OK? Method Status                Protocol
Vlan1                  172.16.0.100    YES manual up                    up
FastEthernet0/1        unassigned      YES unset  up                    up
FastEthernet0/2        unassigned      YES unset  up                    up
..........
```
//显示了各个端口的 IP 地址、状态等简要信息

> **提示**
> 当然也可以在别的 VLAN 端口上配置 IP 地址,二层交换机上只能有一个 VLAN 端口是打开的,三层交换机上可以有多个 VLAN 端口同时打开。有关 VLAN 端口的知识在后续章节详细介绍,这里简单地介绍在 VLAN1 端口上配置管理 IP。

(2) telnet

从 R1 或者 R2 上 Telnet 交换机。

```
R1#telnet 172.16.0.100
Trying 172.16.0.100 ... Open
User Access Verification
Username: test
Password:
S1>
```

(3) SSH

从 R1 或者 R2 上 SSH 到交换机。如果从计算机上 SSH 到交换机,须要安装支持 SSH 协议的软件,例如,SecureCRT 等。

```
R1#ssh -v 2 -c aes128-cbc -l test 172.16.0.100
Password:
S1>
```
路由器上的 SSH 命令参数含义如下:
```
R1#ssh ?
  -c    Select encryption algorithm
  -l    Log in using this user name
  -m    Select HMAC algorithm
  -o    Specify options
  -p    Connect to this port
  -v    Specify SSH Protocol Version
```

```
WORD    IP address or hostname of a remote system
```

（4）监控 MAC 地址表

```
S1#show mac-address-table
          Mac Address Table
-------------------------------------------
Vlan    Mac Address         Type        Ports
----    -----------         --------    -----
All     0100.0ccc.cccc      STATIC      CPU
..............................
All     0180.c200.000f      STATIC      CPU
All     0180.c200.0010      STATIC      CPU
All     ffff.ffff.ffff      STATIC      CPU
 1      0016.475c.5070      DYNAMIC     Fa0/1
 1      0016.9d08.3e60      DYNAMIC     Fa0/2
Total Mac Addresses for this criterion: 22
```

以上显示交换机学习到的 MAC 地址，该命令可以带不同参数以显示特定信息，如下所示：

```
S1#show mac-address-table ?
  address       address keyword                          //显示特定 MAC 地址的信息
  aging-time    aging-time keyword                       //显示 MAC 地址的超时时间
  count         count keyword                            //显示 MAC 地址表的统计信息
  dynamic       dynamic entry type                       //显示动态学习到的 MAC 地址表
  interface     interface keyword                        //显示特定端口的 MAC 地址表
  multicast     multicast info for selected wildcard     //显示特定组播的 MAC 地址表
  static        static entry type                        //显示静态学习到的 MAC 地址表
  vlan          VLAN keyword                             //显示特定 VLAN 的 MAC 地址表

S1#clear mac-address-table dynamic
//清除动态学习到的 MAC 地址表

S1(config)#mac-address-table static 0080.c80c.2996 vlan 1 interface f0/1
//配置静态 MAC 地址表，把特定的 MAC 地址绑定在特定的端口上，通常是出于安全的考虑
```

（5）查看端口的信息

```
S1#show interfaces f0/1 counters
Port            InOctets       InUcastPkts      InMcastPkts      InBcastPkts
Fa0/1            84847             912              57                3
Port            OutOctets      OutUcastPkts     OutMcastPkts     OutBcastPkts
Fa0/1           283102             905             1785              216
//以上显示端口收、发各种数据包的数量


S1#show interfaces f0/1 counters errors
Port         Align-Err      FCS-Err      Xmit-Err       Rcv-Err  UnderSize
Fa0/1            0             0            0              0         0
```

```
Port      Single-Col Multi-Col  Late-Col Excess-Col Carri-Sen    Runts      Giants
Fa0/1         0         0          0        0         0            0          0
```
//以上显示端口收、发各种数据包的错误数量

S1#**show interfaces f0/1 status**
```
Port        Name              Status        Vlan    Duplex   Speed Type
Fa0/1                         connected     1       a-full   a-100 10/100BaseTX
```
//以上显示了端口的状态、双工及速率等

S1#**show interfaces f0/1**
FastEthernet0/1 is up, line protocol is up (connected)
　Hardware is Fast Ethernet, address is 0023.aba4.8303 (bia 0023.aba4.8303)
　MTU 1500 bytes, BW 100000 Kbit, DLY 100 usec,
　　reliability 255/255, txload 1/255, rxload 1/255
　Encapsulation ARPA, loopback not set
　Keepalive set (10 sec)
　Full-duplex, 100Mb/s, media type is 10/100BaseTX
　input flow-control is off, output flow-control is unsupported
　………………………………
　　0 babbles, 0 late collision, 0 deferred
　　0 lost carrier, 0 no carrier, 0 PAUSE output
　　0 output buffer failures, 0 output buffers swapped out
//以上显示了端口的各种信息

2.3 实验 2：划分 VLAN

1. 实验目的

通过本实验可以掌握：
① VLAN 的创建。
② 把交换机端口划分到特定 VLAN 的方法。

2. 实验拓扑

划分 VLAN 实验拓扑如图 2-11 所示。

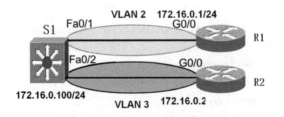

图 2-11　划分 VLAN 的实验拓扑

3. 实验步骤

要配置 VLAN，首先要创建 VLAN，然后把交换机的端口划分到特定的端口上。

（1）把交换机 S1 的配置清空，重启

在交换机 S1 上执行以下命令：

S1#**erase startup-config**
Erasing the nvram filesystem will remove all configuration files! Continue? [confirm]回车
[OK]
//以上命令清除交换机的启动配置
S1#**dir flash:/vlan.dat**
//以上命令清除交换机的 VLAN 文件，如果交换机上没有配置 VLAN 则会提示错误，可忽略错误

在划分 VLAN 前，配置路由器 R1 和 R2 的 G0/0 端口，从 R1 上 ping R2（172.16.0.2）。默认时，交换机的全部端口都在 VLAN1 上，R1 和 R2 应该能够通信。

（2）在 S1 上创建 VLAN

S1(config)#**vlan 2**
S1(config-vlan)#**name VLAN2**
S1(config-vlan)#**exit**
S1(config)#**vlan 3**
S1(config-vlan)#**name VLAN3**

 技术要点

创建 VLAN 时，VLAN 编号的范围如表 2-5 所示。

表 2-5　VLAN 编号的范围

VLAN 范围	使用方式	用　　途	通过 VTP 进行传播？
0,4095	保留	仅限系统使用，用户不能使用和查看	—
1	保留	默认的 VLAN，用户不能创建和删除，但可以使用	是
2,1001	正常	以太网的 VLAN，用户可以创建和删除	是
1002,1005	保留	用于 FDDI、Token Ring，用户不能创建和删除，但可以使用	是
1006,1024	保留	仅限系统使用，用户不能使用和查看	—
1025,4094	扩展	仅用于以太网，用户可以创建和删除，但只能在透明模式下	否

（3）把端口划分到 VLAN 中

S1(config)#**interface fastEthernet0/1**
S1(config-if)#**switchport**
//把该端口配置为二层端口，这是默认值。S1 是三层交换机，端口可以作为三层端口（使用 **no switchport** 命令）

S1(config-if)#**switchport host**
switchport mode will be set to access
spanning-tree portfast will be enabled
channel group will be disabled
//把该端口配置为主机设备所连接的端口。该命令会同时启用 portfast 并禁用 EtherChannel 特性，这些特性在后面章节介绍

S1(config-if)#**switch mode access**
//把交换机端口的模式改为 access 模式,说明该端口是用于连接计算机的,而不是用于 Trunk
S1(config-if)#**switch access vlan 2**
//把端口 Fa0/1 划分到 VLAN2 中
S1(config)#**interface fastEthernet0/2**
S1(config-if)#**switchport**
S1(config-if)#**switchport host**
S1(config-if)#**switch mode access**
S1(config-if)#**switch access vlan 3**

 提示

如果没有先创建 VLAN,就把端口划分到某一 VLAN,则交换机会自动创建一个 VLAN,VLAN 的名字为 VLANXXXX。默认时,所有交换机端口都在 VLAN1 上,VLAN1 是不能删除的、也不能改名。如果有多个端口须要划分到同一 VLAN 下,也可以采用如下方式以节约时间:

S1(config)#**interface range fastEthernet0/2 -3**
S1(config-if-range)#**switch mode access**
S1(config-if-range)#**switch access vlan 2**

如果要删除该 VLAN,使用"**no vlan 2**"命令即可。删除某一 VLAN 后,要记得把该 VLAN 上的端口重新划分到别的 VLAN 上,否则将导致端口的"消失"。

4. 实验调试

(1) show vlan

该命令用来查看 VLAN 的信息。

```
SW1#show vlan
VLAN Name                          Status     Ports
---- ------------------------------ --------- -------------------------------
1    default                        active    Fa0/3, Fa0/4, Fa0/5, Fa0/6
                                              Fa0/7, Fa0/8, Fa0/9, Fa0/10
                                              Fa0/11, Fa0/12, Fa0/13, Fa0/14
                                              Fa0/16, Fa0/17, Fa0/18, Fa0/19
                                              Fa0/20, Fa0/21, Fa0/22, Fa0/23
                                              Fa0/24, Gi0/1, Gi0/2
2    VLAN2                          active    Fa0/1
3    VLAN3                          active    Fa0/2
1002 fddi-default                   act/unsup
1003 token-ring-default             act/unsup
1004 fddinet-default                act/unsup
1005 trnet-default                  act/unsup
```

//以上的第一列列出了 VLAN ID;第二列列出了 VLAN 的名字;第三列列出了 VLAN 的状态,active 或 act 为激活,unsup 为非挂起;第四列列出了本交换机上属于该 VLAN 的端口

```
VLAN Type  SAID       MTU    Parent RingNo BridgeNo Stp  BrdgMode Trans1 Trans2
---- ----- ---------- ------ ------ ------ -------- ---- -------- ------ ------
1    enet  100001     1500   -      -      -        -    -        0      0
```

```
2      enet    100002       1500   -    -    -    -         -    0       0
3      enet    100003       1500   -    -    -    -         -    0       0
1002   fddi    101002       1500   -    -    -    -         -    0       0
1003   tr      101003       1500   -    -    -    -         -    0       0
VLAN Type  SAID          MTU    Parent RingNo BridgeNo   Stp  BrdgMode Trans1 Trans2
---- ----- -----         ------ ------ ------ --------   ---- -------- ------ ------
1004 fdnet 101004        1500   -      -      -          ieee -        0      0
1005 trnet 101005        1500   -      -      -          ibm  -        0      0
```

//以上显示各个 VLAN 的类型及最大传输单元等信息；其他列的信息较少用到，也超出本书范畴，不再详细解析

```
Remote SPAN VLANs
------------------------------------------------------------------------

Primary Secondary Type              Ports
```

（2）**show vlan brief**

该命令用来查看 VLAN 的信息摘要。

```
S1#show vlan brief
VLAN Name                         Status    Ports
---- -------------------------------- --------- -------------------------------
1    default                          active    Fa0/3, Fa0/4, Fa0/5, Fa0/6
                                                Fa0/7, Fa0/8, Fa0/9, Fa0/10
                                                Fa0/11, Fa0/12, Fa0/13, Fa0/14
                                                Fa0/16, Fa0/17, Fa0/18, Fa0/19
                                                Fa0/20, Fa0/21, Fa0/22, Fa0/23
                                                Fa0/24, Gi0/1, Gi0/2
2    VLAN2                            active    Fa0/1
3    VLAN3                            active    Fa0/2
1002 fddi-default                     act/unsup
1003 token-ring-default               act/unsup
1004 fddinet-default                  act/unsup
1005 trnet-default                    act/unsup
```

//以上显示 VLAN 的简要信息，和 "show vlan" 命令相比，该命令少了 VLAN 的类型和最大传输单元等信息

（3）**show vlan summary**

该命令用来查看 VLAN 的汇总信息。

```
S1#show vlan summary
 Number of existing VLANs          : 7
 Number of existing VTP VLANs      : 7
 Number of existing extended VLANs : 0
```

//以上显示 VLAN 的汇总信息，第一行为全部 VLAN 数量，第二行为普通 VLAN（可以通过 VTP 传播的 VLAN）数量，第三行为扩展 VLAN 数量

（4）**show interfaces   fastEthernet0/1 switchport**

该命令用来查看 Fa0/1 端口作为交换端口的有关信息。

```
S1#show interfaces    fastEthernet0/1 switchport
Name: Fa0/1    //端口的名字
Switchport: Enabled    //端口是交换端口
Administrative Mode: static access    //管理员已经配置端口为 access 模式
Operational Mode: static access
//端口当前的模式为 access 模式,和管理员的配置是一样的。有可能管理员配置的是自动协商,而最终
结果为 access
Administrative Trunking Encapsulation: negotiate
//管理员已经配置端口的 Trunk 封装方式为自动模式
Operational Trunking Encapsulation: native
//端口的 Trunk 封装方式为 native 方式,即不对帧进行重新封装
Negotiation of Trunking: Off
//已经关闭 Trunk 的自动协商
Access Mode VLAN: 2 (VLAN2)    //端口属于 VLAN2
Trunking Native Mode VLAN: 1 (default)
//端口的本征 VLAN 是 1,VLAN1 为默认的 Native VLAN
Administrative Native VLAN tagging: enabled
Voice VLAN: none    //本端口没有配置 Voice VLAN
//以下的信息是有关私有 VLAN 的信息,在后面的小节中介绍
Administrative private-vlan host-association: none
Administrative private-vlan mapping: none
Administrative private-vlan Trunk native VLAN: none
Administrative private-vlan Trunk Native VLAN tagging: enabled
Administrative private-vlan Trunk encapsulation: dot1q
Administrative private-vlan Trunk normal VLANs: none
Administrative private-vlan Trunk private VLANs: none
Operational private-vlan: none
Trunking VLANs Enabled: ALL
Pruning VLANs Enabled: 2-1001
Capture Mode Disabled
Capture VLANs Allowed: ALL
Protected: false
Unknown unicast blocked: disabled
Unknown multicast blocked: disabled
Appliance trust: none
```

(5)VLAN 间的通信

由于交换机 S1 上的 Fa0/1 和 Fa0/2 属于不同的 VLAN,在 R1 上 ping 172.16.0.2 应该不能成功。

 提示

在实际的应用中,不同 VLAN 应该属于不同 IP 地址子网。然而本实验中 VLAN2 和 VLAN3 都是 172.16.0.0/24 子网,仅仅是为了实验的目的。

2.4 实验 3：Trunk 配置

1. 实验目的

通过本实验可以掌握：
① 交换机端口 Trunk 的配置方法。
② Native VLAN 的含义和配置。

2. 实验拓扑

Trunk 配置实验拓扑如图 2-12 所示。

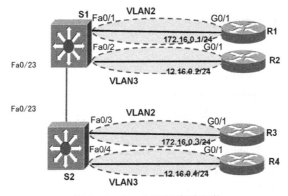

图 2-12　Trunk 配置实验拓扑

3. 实验步骤

（1）在 S1 和 S2 上创建 VLAN 并把端口划分到图 2-12 所示的 VLAN 中

要先关闭不必要的端口，因为在交换机 S1 和 S2 之间还有别的链路存在（见第 1 章图 1-1），以免影响实验，相关信息如下所示：

```
S1(config)#interface fastEthernet0/22
S1(config-if)#shutdown
S1(config-if)#interface fastEthernet0/24
S1(config-if)#shutdown
S2(config)#interface fastEthernet0/22
S2(config-if)#shutdown
S2(config-if)#interface fastEthernet0/24
S2(config-if)#shutdown

S1(config)#vlan 2
S1(config-vlan)#name VLAN2
S1(config-vlan)#vlan 3
S1(config-vlan)#name VLAN3
S1(config-if)#interface fastEthernet0/1
S1(config-if)#switchport mode access
```

```
S1(config-if)#switchport access vlan 2
S1(config-if)#interface fastEthernet0/2
S1(config-if)#switchport mode access
S1(config-if)#switchport access vlan 3
S2(config)#vlan 2
S2(config-vlan)#name VLAN2
S2(config-vlan)#vlan 3
S2(config-vlan)#name VLAN3
S2(config-if)#interface fastEthernet0/3
S2(config-if)#switchport mode access
S2(config-if)#switchport access vlan 2
S2(config-if)#interface fastEthernet0/4
S2(config-if)#switchport mode access
S2(config-if)#switchport access vlan 3
```

（2）配置 Trunk

```
S1(config)#interface fastEthernet0/23
S1(config-if)#switchport Trunk encanpsulation dot1q
```
//配置 Trunk 链路的封装类型，同一链路的两端封装要相同。有的交换机，例如，Catalyst 2950 中有的 IOS 只能封装 dot1q，因此无须执行该命令
```
S1(config-if)#switch mode Trunk     //把端口模式配置为 Trunk
S2(config)#interface fastEthernet0/23
S2(config-if)#switchport Trunk encanpsulation dot1q
S2(config-if)#switch mode Trunk
```

（3）检查 Trunk 链路的状态，测试跨交换机和同一 VLAN 主机间的通信

使用"**show interface fastEthernet0/23 Trunk**"可以查看交换机端口的 Trunk 状态，如下所示：

Port	Mode	Encapsulation	Status	Native vlan
Fa0/23	on	802.1q	Trunking	1

//Fa0/23 端口已经为 Trunk 链路了，封装为 IEEE 802.1q

Port	Vlans allowed on Trunk
Fa0/23	1-4094

//以上显示管理员在 Fa0/23 端口的 Trunk 链路允许 VLAN1~VLAN4094 的数据帧通过。默认允许所有的 VLAN 在 Trunk 链路上通过

Port	Vlans allowed and active in management domain
Fa0/23	1-3

//以上显示 Fa0/23 端口的 Trunk 链路实际允许 VLAN1~VLAN4094 的数据帧通过，之所以和上一段不同，是因为交换机只有 VLAN1、VLAN2 和 VLAN3 存在

Port	Vlans in spanning tree forwarding state and not pruned
Fa0/23	1-3

//以上显示 Fa0/23 端口的 Trunk 链路没有被修剪掉的 VLAN，详细请见第 3 章

需要在链路的两端都确认 Trunk 的形成。测试 R1 和 R3 及 R2 和 R4 之间的通信。由于 R1 和 R3 在同一 VLAN，所以 R1 应该能 ping 通 R3；同样，R2 应该能 ping 通 R4。

```
R1#ping 172.16.0.3
Type escape sequence to abort.
Sending 5, 100-byte ICMP Echos to 172.16.12.3, timeout is 2 seconds:
!!!!!
Success rate is 100 percent (5/5), round-trip min/avg/max = 1/1/4 ms
R2#ping 172.16.0.4
Type escape sequence to abort.
Sending 5, 100-byte ICMP Echos to 172.16.12.4, timeout is 2 seconds:
!!!!!
Success rate is 80 percent (4/5), round-trip min/avg/max = 1/1/4 ms
S1#show interfaces fastEthernet0/23 switchport
Name: Fa0/23
Switchport: Enabled
Administrative Mode: Trunk
Operational Mode: Trunk          //当前端口的模式为 Trunk 模式
Administrative Trunking Encapsulation: dot1q
Operational Trunking Encapsulation: dot1q    //当前端口的 Trunk 封装方式为 dot1q
Negotiation of Trunking: On
Access Mode VLAN: 1 (default)
Trunking Native Mode VLAN: 1 (default)
Administrative Native VLAN tagging: enabled
(省略)
```

（4）配置 Trunk allowed

```
S1(config)#interface fastEthernet0/23
S1(config-if)#switchport Trunk allowed vlan 2,200
//配置 S1 的 Fa0/23 端口的 Trunk 链路只能让 VLAN2 和 VLAN200 的数据通过，这样 R1 应该可以 ping
通 R3，R2 则不可以 ping 通 R4，如下所示：
R1#ping 172.16.0.3
Type escape sequence to abort.
Sending 5, 100-byte ICMP Echos to 172.16.12.3, timeout is 2 seconds:
!!!!!
Success rate is 100 percent (5/5), round-trip min/avg/max = 1/2/4 ms
R2#ping 172.16.0.4
Type escape sequence to abort.
Sending 5, 100-byte ICMP Echos to 172.16.12.4, timeout is 2 seconds:
.....
Success rate is 0 percent (0/5)
S1#show interface fastEthernet0/23 Trunk
```

Port	Mode	Encapsulation	Status	Native vlan
Fa0/23	on	802.1q	Trunking	1

Port	Vlans allowed on Trunk
Fa0/23	2,200

//以上显示管理员在 Fa0/23 端口的 Trunk 链路允许 VLAN2 和 VLAN200 的数据帧通过

Port	Vlans allowed and active in management domain
Fa0/23	2

//以上显示 Fa0/23 端口的 Trunk 链路实际只许 VLAN2 的数据帧通过，之所以和上一段不同是因为交换机上只有 VLAN1、VLAN2 和 VLAN3，而管理员只允许 VLAN2 和 VLAN200 的数据通过

Port	Vlans in spanning tree forwarding state and not pruned
Fa0/23	2

提示

"switchport trunk allowed vlan" 命令有以下选项。

① VLAN ID：VLAN 列表，可以采用 2,3,4-800 这种形式，允许列表中指明的 VLAN 数据通过。

② add：在原有的列表上增加新的 VLAN。

③ all：允许所有的 VLAN。

④ except：除指定 VLAN 以外的 VLAN 都允许通过。

⑤ none：不允许任何 VLAN 通过。

⑥ remove：在原有的列表上删除指定的 VLAN。

（5）配置 Native VLAN

S1(config)#**interface fastEthernet0/23**
S1(config-if)#**switchport trunk native vlan 2**
//在 Trunk 链路上配置 Native VLAN，把它改为 VLAN2，默认是 VLAN1
S2(config)#**interface fastEthernet0/23**
S2(config-if)#**switchport trunk native vlan 2**
S1#**show interface fastEthernet0/23 trunk**

Port	Mode	Encapsulation	Status	Native vlan
Fa0/23	on	802.1q	Trunking	2

//可以查看 Trunk 链路的 Native VLAN 改为 VLAN 2 了

提示

之前介绍在 Trunk 链路上数据帧会根据 ISL 或者 IEEE 802.1q 被重新封装，然而如果是 Native VLAN 的数据，是不会被重新封装就直接在 Trunk 链路上传输的。很显然链路两端的 Native VLAN 是要一样的；如果不一样，交换机会提示出错。

4. 实验调试

下面探讨当两交换机上配置的 Native VLAN 不一样时会出现什么后果。先故意在 Trunk 链路的两端配置不同的 Native VLAN，如下所示：

S1(config)#**interface fastEthernet0/23**
S1(config-if)#**switchport trunk native vlan 2**
S2(config)#**interface fastEthernet0/23**
S2(config-if)#**switchport trunk native vlan 3**

很快，在两个交换机上每隔约一分钟就会出现链路两端的 Native VLAN 不一致的出错提示，如下所示：

　　*Mar　　1 10:52:58.774: %SPANTREE-2-RECV_PVID_ERR: Received BPDU with inconsistent peer vlan id 1 on FastEthernet0/23 VLAN2.
　　*Mar　　1 10:52:58.774: %SPANTREE-2-BLOCK_PVID_LOCAL: Blocking FastEthernet0/23 on VLAN0002. Inconsistent local vlan.

现在在交换机上关闭 STP，STP 是为了防止发生交换环路情况。关于 STP 的详细知识请见本书后面的章节，在图 2-12 的拓扑中不存在交换环路，所以可以关闭 STP，如下所示：

S1(config)#**no spanning-tree vlan 2**
S1(config)#**no spanning-tree vlan 3**
S2(config)#**no spanning-tree vlan 2**
S2(config)#**no spanning-tree vlan 3**

现在从 R1 上 ping R4，如下所示：

R1#**ping 172.16.0.4**
Type escape sequence to abort.
Sending 5, 100-byte ICMP Echos to 172.16.12.4, timeout is 2 seconds:
!!!!!
Success rate is 100 percent (5/5), round-trip min/avg/max = 1/2/4 ms

竟然 ping 通了，然而 R1 和 R4 是在不同的 VLAN 上！ping 通的原因是：由于 S1 的 Fa0/23 端口的 Native VLAN 是 VLAN2，来自 VLAN2 的数据原封不动地从 Fa0/23 端口发送出去到达交换机 S2；S2 的 Fa0/23 端口的 Native VLAN 是 VLAN3，S2 收到了一个没有经过重新封装的数据帧后认为这个帧应该是 VLAN3 的数据帧，发送给了 R4；从 R4 返回到 R1 的数据也是类似的。

2.5　实验 4：DTP 的配置

1. 实验目的

通过本实验可以掌握：
① DTP 的协商规律。
② DTP 的配置。

2. 实验拓扑

DTP 配置实验拓扑如图 2-13 所示。

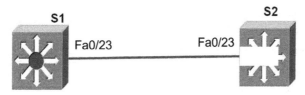

图 2-13　DTP 配置实验拓扑

3. 实验步骤

📖 技术要点

和 DTP 配置有关的命令如下所示，这些命令不能任意组合，须要按规律组合。

① "**switchport trunk encapsulation** { **negotiate** | **isl** | **dot1q** }"：配置 Trunk 链路上的封装类型，可以由双方协商确定，也可以是指定的 isl 或者 dot1q。

② "**switchport nonegotiate**"：Trunk 链路上不发送协商包，默认是发送的。

③ "**switchport mode** { **trunk** | **dynamic desirable** | **dynamic auto** }"：。

- **trunk**：这个设置将端口置为永久 Trunk 模式，封装类型由"switchport trunk encapsulation"命令决定。
- **dynamic desirable**：端口主动变为 Trunk，如果另一端为 negotiate、dynamic desirable 和 dynamic auto，将协商成功。
- **dynamic auto**：被动协商，如果另一端为 negotiate 和 dynamic desirable 将协商成功。

如果想把端口配置为 negotiate，使用：

S1(config-if)#**switchport trunk encapsulation** { **isl** | **dot1q** }
S1(config-if)#**switchport mode trunk**
S1(config-if)#**no switchport nonegotiate**   //这是默认值

如果想把端口配置为 nonegotiate，使用：

S1(config-if)#**switchport trunk encapsulation** { **isl** | **dot1q** }
S1(config-if)#**switchport mode trunk**
S1(config-if)#**switchport nonegotiate**

如果想把端口配置为 desirable，使用：

S1(config-if)#**switchport mode dynamic desirable**
S1(config-if)#**switchport trunk encapsulation** { **negotiate** | **isl** | **dot1q** }

如果想把端口配置为 auto，使用：

S1(config-if)#**switchport mode dynamic auto**
S1(config-if)#**switchport trunk encapsulation** { **negotiate** | **isl** | **dot1q** }

（1）把 S1 的 Fa0/23 端口的 Trunk 配置为 desirable，把 S2 的 Fa0/23 端口的 Trunk 配置为 auto 模式

S1(config)#**interface fastEthernet0/23**
S1(config-if)#**shutdown**
S1(config-if)#**switchport mode dynamic desirable**
S1(config-if)#**switchport trunk encapsulation negotiate**
S1(config-if)#**no shutdown**
S2(config)#**interface fastEthernet0/23**
S2(config-if)#**shutdown**
S2(config-if)#**switchport mode dynamic auto**
S2(config-if)#**switchport trunk encapsulation negotiate**
S2(config-if)#**no shutdown**
S1#**show interfaces fastEthernet0/23 trunk**

| Port | Mode | Encapsulation | Status | Native vlan |

第2章 VLAN、Trunk、VTP与链路聚集

```
Fa0/23        desirable        n-isl        trunking        1
```
//在"Status"列可以看到 Trunk 已经形成,封装为 n-isl,这里的"n"表示封装类型也是自动协商的。需要在两端都进行检查,确认两端都形成 Trunk 才行

```
Port          Vlans allowed on trunk
Fa0/23        1-4094

Port          Vlans allowed and active in management domain
Fa0/23        1-3

Port          Vlans in spanning tree forwarding state and not pruned
Fa0/23        2-3
```

提示

由于交换机有默认配置,在进行以上配置后,使用"**show running**"可能看不到配置的命令。默认时 catalyst 2950 和 3550 的配置是 desirable 模式;而 catalyst 3560 是 auto 模式,所以两台 3560 交换机之间不会自动形成 Trunk,3560 交换机和 2950 交换机之间却可以形成 Trunk。

(2)S1 的 Fa0/23 端口的 Trunk 配置不变,把 S2 的 Fa0/23 端口的 Trunk 配置为 nonegotiate 模式

```
S2(config)#default interface fastEthernet0/23
S2(config)#interface fastEthernet0/23
S2(config-if)#shutdown
S2(config-if)#switchport trunk encapsulation dot1q
S2(config-if)#switchport mode trunk
S2(config-if)#switchport nonegotiate
S2(config-if)#no shutdown
S1#show interfaces fastEthernet0/23 trunk
Port          Mode          Encapsulation    Status          Native vlan
Fa0/23        desirable     negotiate        not-trunking    1
```
//从"Status"列可以看到,并没有形成 Trunk

```
Port          Vlans allowed on trunk
Fa0/23        1

Port          Vlans allowed and active in management domain
Fa0/23        1

Port          Vlans in spanning tree forwarding state and not pruned
Fa0/23        none
```

提示

要确认 Trunk 是否形成,需要同时查看链路两端的信息。经验表明,即使在配置正确的情况下,也有可能不能形成 Trunk,把端口关闭重启即可。

2.6 实验 5：EtherChannel 配置

1. 实验目的

通过本实验可以掌握：
① EtherChannel 的工作原理。
② EtherChannel 的配置。

2. 实验拓扑

EtherChannel 配置实验拓扑如图 2-14 所示。

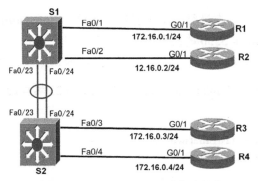

图 2-14 EtherChannel 配置实验拓扑

3. 实验步骤

构成 EtherChannel 的端口必须具有相同的特性：access/trunk 模式、Trunking 的状态、Trunk 的封装方式、双工、速率及所属的 VLAN 等。配置 EtherChannel 有手动配置和自动配置（PAGP 或者 LACP）两种方法。手动配置就是管理员指明哪些端口形成 EtherChannel；自动配置就是让 EtherChannel 协商协议自动协商 EtherChannel 的建立，协商的协议有 PAGP 或者 LAGP 两种。

（1）手动配置 EtherChannel

S1(config)#**interface fastEthernet0/22**
S1(config-if)#**shutdown**
//关闭不必要端口，否则会影响测试
S1(config)#**interface port-channel 1**
//创建以太通道，要指定一个唯一的通道组号，组号的范围是 1～6 的正整数。当要取消 EtherChannel 时用 "**no interface port-channel 1**"
S1(config)#**interface range fastEthernet0/23 -24**
S1(config-if-range)#**channel-group 1 mode on**   //将物理端口指定到已创建的通道中
S1(config-if-range)#**switchport trunk encapsulation dot1q**
S1(config-if-range)#**switchport mode trunk**
//配置通道中的物理端口的属性，这里是配置 Trunk

```
S1(config)# port-channel load-balance dst-ip
```
//配置 EtherChannel 的负载平衡方式,命令格式为 "**port-channel load-balance** *method* ",负载平衡的方式有 dst-ip、dst-mac、src-dst-ip、src-dst-mac、src-ip 和 src-mac。有的交换机还支持 src-prot、dst-port 和 src-dst-port 平衡方式

```
S2(config)#interface port-channel 2
```
//链路两端的 channel-group 是可以不一样的,这个编号只在本地有效
```
S2(config)#interface range fastEthernet0/23 -24
S2(config-if-range)#channel-group 2 mode on
S2(config-if-range)#switchport trunk encapsulation dot1q
S2(config-if-range)#switchport mode trunk
S2(config)#port-channel load-balance dst-ip
```

(2)查看 EtherChannel 信息

```
S1#show etherchannel summary
Flags:   D - down         P - in port-channel
         I - stand-alone  s - suspended
         H - Hot-standby (LACP only)
         R - Layer3       S - Layer2
         U - in use       f - failed to allocate aggregator
         u - unsuitable for bundling
         w - waiting to be aggregated
         d - default port
Number of channel-groups in use: 1
Number of aggregators:           1
Group   Port-channel   Protocol      Ports
------+-------------+------------+-----------------------------------------
1       Po1(SU)        -             Fa0/23(P)   Fa0/24(P)
```
//可以看到编号为 group 1 的 EtherChannel 已经形成,"SU"表示 EtherChannel 正常。如果显示为"SD",先把 port-channel 端口关掉重新开启试试,还不正常的话再查找其他原因。注意:应在链路的两端都进行检查,确认两端都形成以太通道才行

```
S1#show etherchannel load-balance
EtherChannel Load-Balancing Operational State (dst-ip):
Non-IP: Destination MAC address
  IPv4: Destination IP address
  IPv6: Destination IP address
```
//以上显示了 EtherChannel 的负载平衡方式,IPv4 和 IPv6 数据包均基于目的 IP 进行负载平衡,而对于非 IP 包则基于目的 MAC 进行负载平衡

```
S1#show interfaces f0/23 etherchannel
Port state       = Up Mstr In-Bndl
Channel group = 1              Mode = On/FEC           Gcchange = -
Port-channel  = Po1             GC   =  -              Pseudo port-channel = Po1
```

```
Port index        = 0           Load = 0x00            Protocol =    -
Age of the port in the current state: 00d:01h:31m:50s
//可以查看端口在 EtherChannel 中的角色

S1#show etherchannel port-channel
                Channel-group listing:
                ----------------------

Group: 1
----------
                Port-channels in the group:
                ---------------------------

Port-channel: Po1
------------
Age of the Port-channel   = 00d:01h:38m:12s
Logical slot/port    = 2/1           Number of ports = 2
GC                   = 0x00010001    HotStandBy port = null
Port state           = Port-channel Ag-Inuse   //端口的状态
Protocol             =PAgP                     //使用的协商协议
Ports in the Port-channel:

Index   Load   Port        EC state        No of bits
------+------+-------------+---------------+-----------
  0     00    Fa0/23       Desirable-Sl       0
  0     00    Fa0/24       Desirable-Sl       0
//以上列出了该通道包含的端口
Time since last port bundled:      00d:00h:10m:58s     Fa0/24
Time since last port Un-bundled: 00d:00h:14m:33s       Fa0/24


S1#show etherchannel protocol
                Channel-group listing:
                ----------------------

Group: 1
----------
Protocol:   PAgP
//以上显示了各个 Channel-group 使用的协商协议
```

 提示

选择正确的负载平衡方式可以使得负载平衡度更好，假设图 2-14 中的交换机 S2 上接的是服务器，客户计算机接在交换机 S1 上，这时在交换机 S1 上应该配置基于 src-ip 的负载平衡方式，而在 S2 交换机上应该配置基于 dst-ip 的负载平衡方式。

（3）配置 PAGP

 提示

① 要想把端口配置为 PAGP 的 desirable 模式使用命令 "channel-protocol pagp" 和

"channel-group 1 mode desirable"。

② 要想把端口配置为 PAGP 的 auto 模式使用命令"channel-protocol pagp"和"channel-group 1 mode auto"。

这里进行如下配置：

S1(config)#**default interface range fastEthernet0/23 -24**
//把 Fa0/23 和 Fa0/24 端口的配置恢复为出厂状态
S1(config)#**interface range fastEthernet0/23 -24**
S1(config-if-range)#**channel-protocol pagp**
//配置采用 PAGP 协议协商 etherchannel，PAGP 是默认协议，可以不配置
S1(config-if-range)#**channel-group 1 mode desirable**
//配置 PAGP 的模式为 desirable 模式
S1(config-if-range)#**switchport trunk encapsulation dot1q**
S1(config-if-range)#**switchport mode trunk**

S2(config)#**default interface range fastEthernet0/23 -24**
S2(config)#**interface range fastEthernet0/23 -24**
S2(config-if-range)#**channel-protocol pagp**
S2(config-if-range)#**channel-group 2 mode auto**
S2(config-if-range)#**switchport trunk encapsulation dot1q**
S2(config-if-range)#**switchport mode trunk**

S1#**show etherchannel port-channel**
                  Channel-group listing:

Group: 1

                  Port-channels in the group:

Port-channel: Po1

Age of the Port-channel   = 00d:01h:48m:03s
Logical slot/port   = 2/1   Number of ports = 2
GC   = 0x00010001   HotStandBy port = null
Port state   = Port-channel Ag-Inuse
Protocol   = PAgP
Ports in the Port-channel:
Index   Load   Port   EC state   No of bits
------+------+------+------------------+-----------
  0   00   **Fa0/23**   **Desirable-Sl**   0
  0   00   **Fa0/24**   **Desirable-Sl**   0
Time since last port bundled: 00d:00h:00m:36s   Fa0/23
Time since last port Un-bundled: 00d:00h:03m:19s   Fa0/24
//协议是 PAGP，Fa0/23 和 Fa0/24 端口在 group1 中

```
S1#show etherchannel summary
Flags:   D - down          P - bundled in port-channel
         I - stand-alone   s - suspended
         H - Hot-standby (LACP only)
         R - Layer3        S - Layer2
         U - in use        f - failed to allocate aggregator

         M - not in use, minimum links not met
         u - unsuitable for bundling
         w - waiting to be aggregated
         d - default port

Number of channel-groups in use: 1
Number of aggregators:           1

Group  Port-channel  Protocol    Ports
------+-------------+-----------+-----------------------------------------------
1      Po1(SU)        PAgP        Fa0/23(P)    Fa0/24(P)
```
//编号为 group 1 的 EtherChannel 已经形成,"SU"表示 EtherChannel 正常。

(4) 配置 LACP

 提示

① 要想把端口配置为 LACP 的 active 模式使用命令"channel-protocol lacp"和"channel-group 1 mode active"。

② 要想把端口配置为 LACP 的 passive 模式使用命令"channel-protocol lacp"和"channel-group 1 mode passive"。

这里进行如下配置:

```
S1(config)#default interface range fastEthernet0/23 -24
S1(config)#interface range fastEthernet0/23 -24
S1(config-if-range)#channel-protocol lacp
```
//配置采用 LACP 协议协商 EtherChannel
```
S1(config-if-range)#channel-group 1 mode active
```
//配置 LACP 的模式为 auto 模式
```
S1(config-if-range)#switchport trunk encapsulation dot1q
S1(config-if-range)#switchport mode trunk
S1(config-if-range)lacp port-priority 100
```
//配置端口的优先级。例如,如果绑定端口数超过 8 个,由于最多只能有 8 个端口是活动的,这时优先级将决定哪些端口是活动的。优先级数值越小,优先级越高。如果优先级相同,则进一步比较端口的 MAC 地址,值小的优先级高
```
S2(config)#default interface range fastEthernet0/23 -24
S2(config)#interface range fastEthernet0/23 -24
S2(config-if-range)#channel-protocol lacp
```

第 2 章 VLAN、Trunk、VTP 与链路聚集

```
S2(config-if-range)#channel-group 2 mode passive
S2(config-if-range)#switchport trunk encapsulation dot1q
S2(config-if-range)#switchport mode trunk
```

4. 实验调试

（1）show etherchannel summary

该命令用来查看 EtherChannel 的简要信息。

```
S1#show etherchannel summary
Flags:   D - down         P - in port-channel
         I - stand-alone  s - suspended
         H - Hot-standby (LACP only)
         R - Layer3       S - Layer2
         U - in use       f - failed to allocate aggregator

         u - unsuitable for bundling
         w - waiting to be aggregated
         d - default port

Number of channel-groups in use: 1
Number of aggregators:           1
Group  Port-channel  Protocol    Ports
------+-------------+-----------+-----------------------------------------------
1      Po1(SU)       PAgP        Fa0/23(P)   Fa0/24(P)
```
//可以看到 EtherChannel 协商成功。注意：应在链路的两端都进行检查，确认两端都形成以太通道才行

（2）测试负载均衡

在 R1 上分别 ping 172.16.0.3 和 172.16.0.4 一千个数据包，分析数据包路径。通过查看端口上收、发数据包统计数据来分析数据包路径。

首先查看端口上的统计数：

```
S1#show interfaces f0/23 counters
Port            InOctets    InUcastPkts   InMcastPkts   InBcastPkts
Fa0/23          9254410         61791          4266          8133
Port            OutOctets   OutUcastPkts  OutMcastPkts  OutBcastPkts
Fa0/23          1898236          1599           663          3524
S1#show interfaces f0/24 counters
Port            InOctets    InUcastPkts   InMcastPkts   InBcastPkts
Fa0/24          3571473          1585          3956         12830
Port            OutOctets   OutUcastPkts  OutMcastPkts  OutBcastPkts
Fa0/24          7376331         61777           543            68
```

从 R1 ping 172.16.0.3 一千个数据包。

```
R1#ping ip 172.16.0.3 repeat 1000
S1#show interfaces f0/23 counters
Port            InOctets    InUcastPkts   InMcastPkts   InBcastPkts
```

Port				
Fa0/23	9378924	**62793**	4278	8168
Port	OutOctets	OutUcastPkts	OutMcastPkts	OutBcastPkts
Fa0/23	1903176	**1601**	666	3534

S1#show interfaces f0/24 counters

Port	InOctets	InUcastPkts	InMcastPkts	InBcastPkts
Fa0/24	3572071	**1587**	3962	12830
Port	OutOctets	OutUcastPkts	OutMcastPkts	OutBcastPkts
Fa0/24	7494937	**62778**	546	68

可以从 Fa0/23 和 Fa0/24 端口上收、发的数据包统计数据看出：ping 包（到 172.16.0.3）从 Fa0/24 端口发出，到 172.16.0.1 从 Fa0/23 端口返回，如图 2-15 的上图所示。

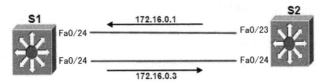

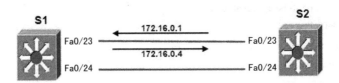

图 2-15 基于 DST-ip 负载均衡结果

同样，从 R1 ping 172.16.0.4 一千个数据包，通过查看端口上收、发数据包统计数据来分析数据包路径：

S1#show interfaces f0/23 counters

Port	InOctets	InUcastPkts	InMcastPkts	InBcastPkts
Fa0/23	9398290	**62799**	4310	8261
Port	OutOctets	OutUcastPkts	OutMcastPkts	OutBcastPkts
Fa0/23	1915904	**1607**	671	3559

S1#show interfaces f0/24 counters

Port	InOctets	InUcastPkts	InMcastPkts	InBcastPkts
Fa0/24	3574091	**1593**	3979	12830
Port	OutOctets	OutUcastPkts	OutMcastPkts	OutBcastPkts
Fa0/24	7497255	**62784**	553	70

R1#**ping ip 172.16.0.4 repeat 1000**
S1#**show interfaces f0/23 counters**

Port	InOctets	InUcastPkts	InMcastPkts	InBcastPkts
Fa0/23	9525758	**63802**	4327	8310
Port	OutOctets	OutUcastPkts	OutMcastPkts	OutBcastPkts
Fa0/23	2040730	**2610**	673	3573

S1#**show interfaces f0/24 counters**

| Port | InOctets | InUcastPkts | InMcastPkts | InBcastPkts |

Fa0/24	3575293	**1596**	3988	12830
Port	OutOctets	OutUcastPkts	OutMcastPkts	OutBcastPkts
Fa0/24	7497975	**62787**	556	70

可以从 Fa0/23 和 Fa0/24 端口上收、发数据包统计数据看出：ping 包（到 172.16.0.4）从 Fa0/23 端口发出，到 172.16.0.1 从 Fa0/23 端口返回，如图 2-15 的下图所示。

（3）测试 Etherchannel 的冗余功能

从 R1 连续 ping R3（172.16.0.3），命令为"**ping 172.16.0.3 repeat 9000000**"。在 S1 上关闭 Fa0/24，在 R1 上查看通信应该不中断。

```
S1#show etherchannel summary
Flags:  D - down         P - in port-channel
        I - stand-alone  s - suspended
        H - Hot-standby (LACP only)
        R - Layer3       S - Layer2
        U - in use       f - failed to allocate aggregator
        u - unsuitable for bundling
        w - waiting to be aggregated
        d - default port
Number of channel-groups in use: 1
Number of aggregators:           1

Group  Port-channel  Protocol    Ports
------+-------------+-----------+-----------------------------------------------
1      Po1(SU)         -         Fa0/23(P)   Fa0/24(D)
// S1 交换机上的 Fa0/24 端口关闭后，EtherChannel group 只有一个端口了，Fa0/24 处于 down 状态。这时通过 Fa0/24 发往 172.16.0.3 的流量会自动转移到 Fa0/23 端口上
```

2.7　实验 6：VTP 配置

1. 实验目的

通过本实验可以掌握：
① VTP 三种模式的区别。
② VTP 的配置。

2. 实验拓扑

VTP 配置实验拓扑如图 2-16 所示。

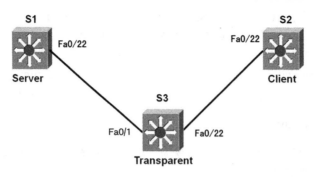

图 2-16 VTP 配置实验拓扑

3. 实验步骤

（1）把 3 台交换机的配置清除干净，重启交换机

```
S1#delete flash:vlan.dat
S1#erase startup-config
S1#reload
S2#delete flash:vlan.dat
S2#erase startup-config
S2#reload
S3#delete flash:vlan.dat
S3#erase startup-config
S3#reload
```

提示

重启 3 台交换机的时间不能相差太大，要保证任何交换机重启之前另一交换机没有启动完毕，避免重新启动的交换机从未重新启动的交换机学到旧 VTP 信息。

（2）关闭不必要的端口以免影响实验，配置 S1 和 S3 之间及 S2 和 S3 之间链路的 Trunk

```
S1(config)#interface fastEthernet0/23
S1(config-if)#shutdown
S1(config-if)#interface fastEthernet0/24
S1(config-if)#shutdown
S1(config)#interface fastEthernet0/22
S1(config-if)#switchport mode dynamic desirable

S2(config)#interface fastEthernet0/23
S2(config-if)#shutdown
S2(config)#interface fastEthernet0/24
S2(config-if)#shutdown
S2(config)#interface fastEthernet0/22
S2(config-if)#switchport mode dynamic desirable

S3(config)#interface fastEthernet0/1
```

```
S3(config-if)#switchport mode dynamic desirable
S3(config)#interface fastEthernet0/2
S3(config-if)#switchport mode dynamic desirable
S3#show interfaces trunk
Port        Mode         Encapsulation    Status      Native vlan
Fa0/1       desirable    802.1q           trunking    1
Fa0/2       desirable    802.1q           trunking    1
（省略）
//可以看到 Trunk 已经正常，由于 VTP 是在 Trunk 链路上传输的，因此 Trunk 非常重要
```

（3）检查默认的 VTP 情况

```
S1#show vtp status
VTP Version capable             : 1 to 3            //该 VTP 支持版本 1～3
VTP version running             : 1                 //当前运行的是版本 1
VTP Domain Name                 :                   //当前 VTP 域名为空
VTP Pruning Mode                : Disabled          //当前 VTP 没有启用修剪
VTP Traps Generation            : Disabled          //没有把 VTP 提示信息等发送到 SNMP 服务器
Device ID                       : d0c7.89ab.1180    //该设备的 MAC 地址
Configuration last modified by 0.0.0.0 at 0-0-00 00:00:00
//VTP 配置的修改者的 IP 地址（即是在哪个交换机上修改的）和修改时间
Local updater ID is 0.0.0.0 (no valid interface found)    //更新者的 ID


Feature VLAN:
--------------
VTP Operating Mode              : Server            //当前模式是 Server
Maximum VLANs supported locally : 1005              //VTP 支持的最大 VLAN 数量
Number of existing VLANs        : 5                 //当前交换机上有 5 个 VLAN
Configuration Revision          : 0                 //默认修订号为 0，该数值非常重要
MD5 digest                      : 0x57 0xCD 0x40 0x65 0x63 0x59 0x47 0xBD
                                  0x56 0x9D 0x4A 0x3E 0xA5 0x69 0x35 0xBC    //MD5 值
```

同样，S2 和 S3 交换的 VTP 模式也都为 Server 模式。

（4）配置 S1 为 VTP Server，配置 VTP 域名，创建 VLAN2

```
S1(config)#vtp mode server    //配置 S1 为 VTP Server，实际上这是默认值
Device mode already VTP SERVER.
S1(config)#vtp domain VTP-TEST    //配置 VTP 域名，默认为空
Changing VTP domain name from NULL to VTP-TEST


S1#show vtp status
VTP Version capable             : 1 to 3
VTP version running             : 1
VTP Domain Name                 : VTP-TEST    //可以看到 VTP 的域名已经更改了
VTP Pruning Mode                : Disabled
```

```
VTP Traps Generation              : Disabled
Device ID                         : d0c7.89ab.1180
Configuration last modified by 0.0.0.0 at 0-0-00 00:00:00
Local updater ID is 0.0.0.0 (no valid interface found)

Feature VLAN:
--------------
VTP Operating Mode                : Server
Maximum VLANs supported locally   : 1005
Number of existing VLANs          : 5
Configuration Revision            : 0
MD5 digest                        : 0xB1 0x8E 0x7C 0xA5 0xD8 0x67 0x3C 0xAE
                                    0x97 0x14 0x96 0x15 0xEE 0xD6 0xFB 0x92

S1(config)#vlan 2
S1(config-vlan)#name VLAN2-TEST    //在 S1 上创建 VLAN2
S1#show vlan
VLAN Name                         Status    Ports
---- -------------------------------- --------- -------------------------------
1    default                          active    Fa0/1, Fa0/2, Fa0/3, Fa0/4
                                                Fa0/5, Fa0/6, Fa0/7, Fa0/8
                                                Fa0/9, Fa0/10, Fa0/11, Fa0/12
                                                Fa0/13, Fa0/14, Fa0/16, Fa0/17
                                                Fa0/18, Fa0/19, Fa0/20, Fa0/21
                                                Fa0/22, Fa0/23, Fa0/24, Gi0/1
                                                Gi0/2
2    VLAN2-TEST                       active    //刚刚创建的 VLAN
S1#show vtp status
......... .........
Feature VLAN:
--------------
VTP Operating Mode                : Server
Maximum VLANs supported locally   : 1005
Number of existing VLANs          : 6        //VLAN 数量变为 6 个
Configuration Revision            : 1        //每当 VLAN 信息有变化时,修订号会增加 1
MD5 digest                        : 0x67 0xAE 0xFC 0x2C 0x63 0xD4 0x36 0x70
                                    0x70 0xD3 0x9D 0x58 0x59 0xA1 0xA1 0xED
```

在交换机 S2 和 S3 上进行检查:

```
S3#show vtp status
VTP Version capable               : 1 to 3
VTP version running               : 1
VTP Domain Name                   : VTP-TEST      //VTP 的域名也变为 VTP-TEST
VTP Pruning Mode                  : Disabled
VTP Traps Generation              : Disabled
```

Device ID   : d0c7.89c2.8380
Configuration last modified by 0.0.0.0 at 3-1-93 00:12:36
Local updater ID is 0.0.0.0 (no valid interface found)

Feature VLAN:

VTP Operating Mode   : Server
Maximum VLANs supported locally : 1005
Number of existing VLANs   : 6   //VLAN 数量也变为 6 个
Configuration Revision   : 1
MD5 digest   : 0x67 0xAE 0xFC 0x2C 0x63 0xD4 0x36 0x70
                                  0x70 0xD3 0x9D 0x58 0x59 0xA1 0xA1 0xED

S3#**show vlan**
VLAN Name   Status   Ports
---- -------------------------------- ---------- --------------------------------
1   default   active   Fa0/3, Fa0/4, Fa0/5, Fa0/6
                                                      Fa0/7, Fa0/8, Fa0/9, Fa0/10
                                                      Fa0/11, Fa0/12
2   **VLAN2-TEST**   **active**   //自动学习到了在交换机 S1 上创建的 VLAN

S2#**show vtp status**
VTP Version capable   : 1 to 3
VTP version running   : 1
VTP Domain Name   : **VTP-TEST**   //VTP 的域名也变为 VTP-TEST
VTP Pruning Mode   : Disabled
VTP Traps Generation   : Disabled
Device ID   : d0c7.89c2.3100
Configuration last modified by 0.0.0.0 at 3-1-93 00:12:36
Local updater ID is 0.0.0.0 (no valid interface found)

Feature VLAN:

VTP Operating Mode   : Server
Maximum VLANs supported locally : 1005
Number of existing VLANs   : 6   //VLAN 数量也变为 6 个
Configuration Revision   : 1
MD5 digest   : 0x67 0xAE 0xFC 0x2C 0x63 0xD4 0x36 0x70
                                  0x70 0xD3 0x9D 0x58 0x59 0xA1 0xA1 0xED

S2#**show vlan**
VLAN Name   Status   Ports
---- -------------------------------- ---------- --------------------------------

1	default		active	Fa0/1, Fa0/2, Fa0/3, Fa0/4
				Fa0/5, Fa0/6, Fa0/7, Fa0/8
				Fa0/9, Fa0/10, Fa0/11, Fa0/12
				Fa0/13, Fa0/14, Fa0/16, Fa0/17
				Fa0/18, Fa0/19, Fa0/20, Fa0/21
				Fa0/22, Fa0/23, Fa0/24, Gi0/1
				Gi0/2
2	VLAN2-TEST		active	//自动学习到了在交换机 S1 上创建的 VLAN

📁 提示

在以上配置中，只在 S1 上配置了 VTP 域名，而在 S3 和 S2 上并没有进行任何 VTP 配置。当交换机的 VTP 域名为空时，如果它收到的 VTP 通告中带有域名，该交换机将把 VTP 域名自动更改为 VTP 通告中的域名，所以没有 VTP 域名的交换机能从邻居处自动学习到 VTP 域名。在以上实验中，在 S1 上配置 VTP 域名并创建 VLAN，这就导致 S1 发送 VTP 通告，S3 和 S2 交换机就自动学习到 VTP 域名。一旦 VTP 域名不为空，交换机就不会学习域名了，也就是说如果在 S1 上再次修改 VTP 域名，S2 和 S3 交换机的 VTP 域名不会跟着再更新了。

（5）配置 S3 为 VTP transparent

```
S3(config)#vtp mode transparent
//把 S3 的 VTP 模式改为 transparent
S3#show vtp status
VTP Version capable             : 1 to 3
VTP version running             : 1
VTP Domain Name                 : VTP-TEST
VTP Pruning Mode                : Disabled
VTP Traps Generation            : Disabled
Device ID                       : d0c7.89c2.8380
Configuration last modified by 0.0.0.0 at 3-1-93 00:12:36

Feature VLAN:
--------------
VTP Operating Mode              : Transparent   //VTP 为 transparent 模式
Maximum VLANs supported locally : 1005
Number of existing VLANs        : 6
Configuration Revision          : 0         //当为 transparent 模式时，修订号始终为 0
MD5 digest                      : 0x67 0xAE 0xFC 0x2C 0x63 0xD4 0x36 0x70
                                  0x70 0xD3 0x9D 0x58 0x59 0xA1 0xA1 0xED
```

在 S3 交换机上创建 VLAN3，检查 S1 和 S2 上的 VLAN 信息，如下所示：

```
S3(config)#vlan 3
S3(config-vlan)#name TEST-VLAN3
S3#show vlan
VLAN Name                                Status    Ports
```

```
---- -------------------------------  ---------  -------------------------------
1    default                          active     Fa0/3, Fa0/4, Fa0/5, Fa0/6
                                                 Fa0/7, Fa0/8, Fa0/9, Fa0/10
                                                 Fa0/11, Fa0/12
2    VLAN2-TEST                       active
3    TEST-VLAN3                       active     //新创建的 VLAN 3
```

S1#**show vlan**

```
VLAN Name                             Status     Ports
---- -------------------------------  ---------  -------------------------------
1    default                          active     Fa0/1, Fa0/2, Fa0/3, Fa0/4
                                                 Fa0/5, Fa0/6, Fa0/7, Fa0/8
                                                 Fa0/9, Fa0/10, Fa0/11, Fa0/12
                                                 Fa0/13, Fa0/14, Fa0/16, Fa0/17
                                                 Fa0/18, Fa0/19, Fa0/20, Fa0/21
                                                 Fa0/22, Fa0/23, Fa0/24, Gi0/1
                                                 Gi0/2
2    VLAN2-TEST                       active
```

//在 S1 上并没有学习到 VLAN3，S2 上也是如此。以上表明，在 Transparent 交换机上可以创建 VLAN，然而这些 VLAN 信息并不会通告出去，仅本地有效

在 S2 交换机上创建 VLAN4，检查 S1 和 S2 上的 VLAN 信息，如下所示：

S2(config)#**vlan 4**
S2(config-vlan)#**name TEST-VLAN4**
S2#**show vlan**

```
VLAN Name                             Status     Ports
---- -------------------------------  ---------  -------------------------------
1    default                          active     Fa0/1, Fa0/2, Fa0/3, Fa0/4
                                                 Fa0/5, Fa0/6, Fa0/7, Fa0/8
                                                 Fa0/9, Fa0/10, Fa0/11, Fa0/12
                                                 Fa0/13, Fa0/14, Fa0/16, Fa0/17
                                                 Fa0/18, Fa0/19, Fa0/20, Fa0/21
                                                 Fa0/22, Fa0/23, Fa0/24, Gi0/1
                                                 Gi0/2
2    VLAN2-TEST                       active
4    TEST-VLAN4                       active     //新创建的 VLAN4
```

S1#**show vlan**

```
VLAN Name                             Status     Ports
---- -------------------------------  ---------  -------------------------------
1    default                          active     Fa0/1, Fa0/2, Fa0/3, Fa0/4
                                                 Fa0/5, Fa0/6, Fa0/7, Fa0/8
                                                 Fa0/9, Fa0/10, Fa0/11, Fa0/12
                                                 Fa0/13, Fa0/14, Fa0/16, Fa0/17
                                                 Fa0/18, Fa0/19, Fa0/20, Fa0/21
                                                 Fa0/22, Fa0/23, Fa0/24, Gi0/1
                                                 Gi0/2
```

| 2 | VLAN2-TEST | | active | |
| 4 | TEST-VLAN4 | | active | //S1 自动学习到在 S2 上新创建的 VLAN4 |

S3#**show vlan**

VLAN	Name	Status	Ports
1	default	active	Fa0/3, Fa0/4, Fa0/5, Fa0/6
			Fa0/7, Fa0/8, Fa0/9, Fa0/10
			Fa0/11, Fa0/12
2	VLAN2-TEST	active	
3	TEST-VLAN3	active	//S3 上并没有新创建的 VLAN4

//以上表明，在一个 VTP 域中可以有多个 VTP Server，在任何一个 VTP Server 上都可以创建和修改 VLAN 信息并通告到其他交换机上。Transparent 交换机会转发 VTP 通告，但是并不会根据 VTP 通告更新自己的 VLAN 信息

（6）配置 S2 和 S3 为 VTP Client

S3(config)#**vtp mode client**
Setting device to VTP CLIENT mode.
S2(config)#**vtp mode client**
Setting device to VTP CLIENT mode.

S2#**show vtp status**
………………
Feature VLAN:

VTP Operating Mode	: Client	//VTP 模式改为 Client 了
Maximum VLANs supported locally	: 1005	
Number of existing VLANs	: 7	
Configuration Revision	: 2	
MD5 digest	: 0x4F 0x80 0xAE 0x1F 0x38 0xCB 0xEF 0x47	
	0xD3 0xB1 0x8B 0x3E 0xC6 0xE2 0x49 0x3D	

S2(config)#**vlan 5**
VTP VLAN configuration not allowed when device is in CLIENT mode.
//以上表明，在 Client 模式交换机上不能创建和修改 VLAN 信息

在交换机 S1 上创建 VLAN6，检查 S2 和 S3 上的 VLAN 信息，如下所示：

S1(config)#**vlan 6**
S1(config-vlan)#**name TEST-VLAN6**
S2#**show vlan**

VLAN	Name	Status	Ports
1	default	active	Fa0/1, Fa0/2, Fa0/3, Fa0/4
			Fa0/5, Fa0/6, Fa0/7, Fa0/8
			Fa0/9, Fa0/10, Fa0/11, Fa0/12
			Fa0/13, Fa0/14, Fa0/16, Fa0/17

第 2 章　VLAN、Trunk、VTP 与链路聚集

```
                                        Fa0/18, Fa0/19, Fa0/20, Fa0/21
                                        Fa0/22, Fa0/23, Fa0/24, Gi0/1
                                        Gi0/2
2    VLAN2-TEST                 active
4    TEST-VLAN4                 active
6    TEST-VLAN6                 active        //自动学习到 VLAN 6
S3#show vlan
VLAN Name                       Status   Ports
---- ---------------------------- --------- ------------------------------
1    default                    active   Fa0/3, Fa0/4, Fa0/5, Fa0/6
                                         Fa0/7, Fa0/8, Fa0/9, Fa0/10
                                         Fa0/11, Fa0/12
2    VLAN2-TEST                 active
4    TEST-VLAN4                 active
6    TEST-VLAN6                 active        //自动学习到 VLAN 6
//以上表明，在 Client 模式的交换机上不仅可以转发 VTP 通告，并且还会根据 VTP 通告更新自己的
VLAN 信息
```

（7）配置 VTP 密码

```
S1(config)#vtp password cisco
Setting device VLAN database password to cisco
S2(config)#vtp password cisco
S3(config)#vtp password cisco
//配置 VTP 的密码，目的是为了安全，防止不明身份的交换机加入域中破坏 VLAN 信息，密码是大小
写敏感的
S1#show vtp password      //显示 VTP 的密码
VTP Password: cisco
```

（8）配置 VTP 版本，只能在 Server 上配置

```
S1(config)#vtp version 2
S1#show vtp status
VTP Version capable             : 1 to 3
VTP version running             : 2              //VTP 的 Version2 已经启用
VTP Domain Name                 : VTP-TEST
VTP Pruning Mode                : Disabled
VTP Traps Generation            : Disabled
Device ID                       : d0c7.89ab.1180
Configuration last modified by 0.0.0.0 at 3-1-93 00:29:20
Local updater ID is 0.0.0.0 (no valid interface found)

Feature VLAN:
--------------
VTP Operating Mode              : Server
Maximum VLANs supported locally : 1005
```

Number of existing VLANs	: 8
Configuration Revision	: 4
MD5 digest	: 0xA7 0xE0 0xF1 0xD0 0xA7 0x92 0x0A 0x1C
	0x46 0x24 0xF2 0x7F 0xAC 0x77 0xB5 0x14

S2#**show vtp status**

VTP Version capable	: 1 to 3	
VTP version running	**: 2**	//VTP 的 Version2 已经启用
VTP Domain Name	: VTP-TEST	
VTP Pruning Mode	: Disabled	
VTP Traps Generation	: Disabled	
Device ID	: d0c7.89c2.3100	
Configuration last modified by 0.0.0.0 at 3-1-93 00:29:20		

Feature VLAN:

VTP Operating Mode	: Client
Maximum VLANs supported locally	: 1005
Number of existing VLANs	: 8
Configuration Revision	: 4
MD5 digest	: 0xA7 0xE0 0xF1 0xD0 0xA7 0x92 0x0A 0x1C
	0x46 0x24 0xF2 0x7F 0xAC 0x77 0xB5 0x14

//以上表明，VTP Version2 只需要在任一 VTP Server 上启用，其他交换机会自动启用该功能。VTP Version1 和 VTP Version2 只在一些细微的功能上有区别，因此没有什么实际意义

（9）VTP 域名问题

把 S3 的 VTP 域名改为另一域名，如图 2-17 所示，VTP 模式也恢复为图中所示的模式，完整配置如下：

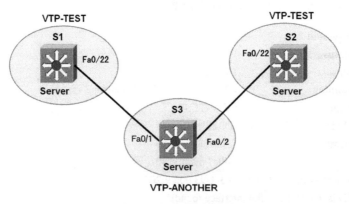

图 2-17　VTP 域名问题

S1(config)#**vtp mode server**
S1(config)#**vtp domain VTP-TEST**
S2(config)#**vtp mode server**

S2(config)#**vtp domain VTP-TEST**

S3(config)#**vtp mode server**

S3(config)#**vtp domain VTP-ANOTHER**

在各交换机上，有可能会看到如下提示：

01:36:38: %DTP-5-DOMAINMISMATCH: Unable to perform trunk negotiation on port Fa0/2 because of VTP domain mismatch.

由于相邻的交换机处于不同的 VTP 域中（即域名不同），将导致交换机之间的链路 Trunk 协商失败（之前已经配置了自动协商），把交换机的端口关闭重启后，就能发现这些链路并不能形成 Trunk，如下所示：

S3#**show interfaces fastEthernet0/1 switchport**
Name: Fa0/1
Switchport: Enabled
Administrative Mode: dynamic desirable
Operational Mode: static access
S3#**show interfaces fastEthernet0/2 switchport**
Name: Fa0/2
Switchport: Enabled
Administrative Mode: dynamic desirable
Operational Mode: static access
//以上表明端口是 Access 模式，Trunk 协商失败。由于 VTP 基于 Trunk 进行工作，显然如果在 S1 上修改 VLAN 信息，S3 和 S2 将无法学习到

手工配置 Trunk 链路，观察 VTP 通告能否跨越不同 VTP 域，如下所示：

S1(config)#**interface fastEthernet0/22**
S1(config-if)#**switchport trunk encapsulation dot1q**
S1(config-if)#**switchport mode trunk**
S2(config)#**interface fastEthernet0/22**
S2(config-if)#**switchport trunk encapsulation dot1q**
S2(config-if)#**switchport mode trunk**
S3(config)#**interface fastEthernet0/1**
S3(config-if)#**switchport trunk encapsulation dot1q**
S3(config-if)#**switchport mode trunk**
S3(config)#**interface fastEthernet0/2**
S3(config-if)#**switchport trunk encapsulation dot1q**
S3(config-if)#**switchport mode trunk**
S1(config)#**vlan 100**
S1(config-vlan)#**name TEST-VLAN100**
S1#**show vlan**

VLAN	Name	Status	Ports
1	default	active	Fa0/1, Fa0/2, Fa0/3, Fa0/4
			Fa0/5, Fa0/6, Fa0/7, Fa0/8
			Fa0/9, Fa0/10, Fa0/11, Fa0/12
			Fa0/13, Fa0/14, Fa0/16, Fa0/17

			Fa0/18, Fa0/19, Fa0/20, Fa0/21
			Fa0/22, Fa0/23, Fa0/24, Gi0/1
			Gi0/2
2	VLAN2-TEST	active	
4	TEST-VLAN4	active	
6	TEST-VLAN6	active	
100	TEST-VLAN100	active	//新创建的 VLAN

```
S3#show vlan
VLAN Name                          Status     Ports
---- -------------------------------- --------- -------------------------------
1    default                          active    Fa0/3, Fa0/4, Fa0/5, Fa0/6
                                                Fa0/7, Fa0/8, Fa0/9, Fa0/10
                                                Fa0/11, Fa0/12
2    VLAN2-TEST                       active
4    TEST-VLAN4                       active
6    TEST-VLAN6                       active
```
//在交换机 S3 和 S2 上并没有 VLAN100，说明不同的 VTP 域之间是不能共享 VLAN 信息的

2.8 实验 7：VTP 覆盖

1. 实验目的

通过本实验可以掌握：
① VTP 通告修订号的作用。
② 在网络中增添交换机的注意事项。

2. 实验拓扑

实验拓扑如图 2-18 所示。

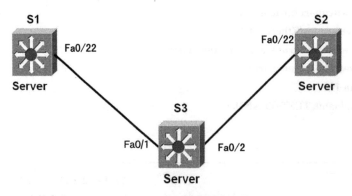

图 2-18　VLAN 信息覆盖实验

第 2 章 VLAN、Trunk、VTP 与链路聚集

3. 实验步骤

VTP 通告中的修订号在 VLAN 信息更新中有很重要的作用，当交换机收到修订号大的通告后，会更新自己的 VLAN 信息；然而如果收到修订号小的通告，将会用自身的 VLAN 信息去更新对方的信息。本节将用 S2 的 VLAN 信息反向更新 S1 和 S3 的信息。

（1）把 3 台交换机的配置清除干净，重启交换机

```
S1#delete flash:vlan.dat
S1#erase startup-config
S1#reload
S2#delete flash:vlan.dat
S2#erase startup-config
S2#reload
S3#delete flash:vlan.dat
S3#erase startup-config
S3#reload
```

（2）关闭不必要的端口，配置 S1 和 S3 之间及 S3 和 S2 之间链路的 Trunk

```
S1(config)#interface fastEthernet0/23
S1(config-if)#shutdown
S1(config-if)#interface fastEthernet0/24
S1(config-if)#shutdown
S1(config)#interface fastEthernet0/22
S1(config-if)#switchport mode dynamic desirable

S2(config)#interface fastEthernet0/23
S2(config-if)#shutdown
S2(config)#interface fastEthernet0/24
S2(config-if)#shutdown
S2(config)#interface fastEthernet0/22
S2(config-if)#switchport mode dynamic desirable

S3(config)#interface fastEthernet0/1
S3(config-if)#switchport mode dynamic desirable
S3(config)#interface fastEthernet0/2
S3(config-if)#switchport mode dynamic desirable
```

（3）按照图 2-18 中所示配置 VTP

```
S1(config)#vtp mode server
S1(config)#vtp domain VTP-TEST
S2(config)#vtp mode server
S2(config)#vtp domain VTP-TEST
S3(config)#vtp mode server
S3(config)#vtp domain VTP-TEST
```

（4）VTP 覆盖问题

关闭 S3 和 S2 的链路，在两交换机上各自修改 VLAN 信息，如下所示：

S3(config)#**interface fastEthernet0/2**
S3(config-if)#**shutdown**
//切断和 S2 连接的链路，目的是把 S2 孤立出来
S3(config)#**vlan 2**
S3(config)#**vlan 3**
S3(config)#**vlan 4**
S3(config)#**vlan 5**
//在 S3 上创建多个 VLAN，用于观察

S3#**show vlan**

VLAN	Name	Status	Ports
1	default	active	Fa0/2, Fa0/3, Fa0/4, Fa0/5
			Fa0/6, Fa0/7, Fa0/8, Fa0/9
			Fa0/10, Fa0/11, Fa0/12
2	VLAN0002	active	
3	VLAN0003	active	
4	VLAN0004	active	
5	VLAN0005	active	

//以上表明 VLAN 创建成功

1002	fddi-default	act/unsup
1003	token-ring-default	act/unsup
1004	fddinet-default	act/unsup
1005	trnet-default	act/unsup

S3#**show vtp status**

```
VTP Version capable              : 1 to 3
VTP version running              : 1
VTP Domain Name                  : VTP-TEST
VTP Pruning Mode                 : Disabled
VTP Traps Generation             : Disabled
Device ID                        : d0c7.89c2.8380
Configuration last modified by 0.0.0.0 at 3-1-93 00:05:39
Local updater ID is 0.0.0.0 (no valid interface found)
Feature VLAN:
--------------
VTP Operating Mode               : Server
Maximum VLANs supported locally  : 1005
Number of existing VLANs         : 9
Configuration Revision           : 4          //VTP 修订为 4，是因为这里创建了多个 VLAN
MD5 digest                       : 0xD6 0x5B 0x4F 0xC7 0x12 0x9F 0xB5 0x27
                                   0xAA 0xAF 0x91 0x74 0x62 0x3D 0x34 0x7D
```

```
S2(config)#vlan 2
S2(config-vlan)#exit
S2(config)#no vlan 2
S2(config)#vlan 2
S2(config-vlan)#exit
S2(config)#no vlan 2
S2(config)#vlan 2
S2(config-vlan)#exit
S2(config)#no vlan 2
//反复执行创建和删除 VLAN 的命令，目的只是为了把 VTP 的修订号变大

S2#show vtp status
VTP Version capable             : 1 to 3
VTP version running             : 1
VTP Domain Name                 : VTP-TEST
VTP Pruning Mode                : Disabled
VTP Traps Generation            : Disabled
Device ID                       : d0c7.89c2.3100
Configuration last modified by 0.0.0.0 at 3-1-93 00:07:29
Local updater ID is 0.0.0.0 (no valid interface found)

Feature VLAN:
--------------
VTP Operating Mode              : Server
Maximum VLANs supported locally : 1005
Number of existing VLANs        : 5
Configuration Revision          : 6
//修订号为 6 了，请确保它比 S3 上的修订号大，否则多执行几次修改 VLAN 的命令
MD5 digest                      : 0x15 0xBE 0xB2 0xB6 0x69 0xC5 0x77 0xFB
                                  0xC0 0x29 0xDF 0x4D 0x52 0x7C 0x69 0x7B

S2#show vlan
VLAN Name                        Status     Ports
---- -------------------------- ---------- -------------------------------
1    default                    active     Fa0/1, Fa0/2, Fa0/3, Fa0/4
                                           Fa0/5, Fa0/6, Fa0/7, Fa0/8
                                           Fa0/9, Fa0/10, Fa0/11, Fa0/12
                                           Fa0/13, Fa0/14, Fa0/15, Fa0/16
                                           Fa0/17, Fa0/18, Fa0/19, Fa0/20
                                           Fa0/21, Fa0/22, Fa0/23, Fa0/24
                                           Gi0/1, Gi0/2
1002 fddi-default               act/unsup
1003 token-ring-default         act/unsup
1004 fddinet-default            act/unsup
```

1005 trnet-default   act/unsup

//以上显示，S2 上除了默认的 VLAN 并无其他 VLAN

打开 S3 和 S2 的链路，观察 VLAN 信息的覆盖情况，如下所示：

S3(config)#**interface fastEthernet0/2**
S3(config-if)#**no shutdown**

等待 30 秒左右，确认 S3 和 S2 之间的链路已经打开，继续：

S3#**show vtp status**

VTP Version capable	: 1 to 3
VTP version running	: 1
VTP Domain Name	: VTP-TEST
VTP Pruning Mode	: Disabled
VTP Traps Generation	: Disabled
Device ID	: d0c7.89c2.8380

Configuration last modified by 0.0.0.0 at 3-1-93 00:07:29
Local updater ID is 0.0.0.0 (no valid interface found)
Feature VLAN:

VTP Operating Mode	: Server
Maximum VLANs supported locally	: 1005
Number of existing VLANs	: 5
Configuration Revision	**: 6**

//VTP 修订号已经变成和 S2 交换机的一样了，因为 S2 的修订号大

MD5 digest   : 0x15 0xBE 0xB2 0xB6 0x69 0xC5 0x77 0xFB
                                0xC0 0x29 0xDF 0x4D 0x52 0x7C 0x69 0x7B

S3#**show vlan**

VLAN	Name	Status	Ports
1	default	active	Fa0/3, Fa0/4, Fa0/5, Fa0/6
			Fa0/7, Fa0/8, Fa0/9, Fa0/10
			Fa0/11, Fa0/12
1002	fddi-default	act/unsup	
1003	token-ring-default	act/unsup	
1004	fddinet-default	act/unsup	
1005	trnet-default	act/unsup	

//以上表明，S3 上除了默认 VLAN，所有的 VLAN 均消失了。因为它的 VLAN 信息和 S2 的信息同步了

📂 **提示**

在准备向网络中添加非刚出厂的交换机时，请先参照本节的步骤（1）把交换机的配置清除干净，否则可能导致和本实验一样的结果，如果因此删除了现有网络的 VLAN，将会使得网络中断。本实验也说明了另一问题，黑客很容易在现有网络中接入一台交换机或者用软件模拟一台交换机，通告大修订号的 VTP 通告，从而破坏网络。可以配置 VTP 密码防止黑客的破坏。

2.9 实验 8：私有 VLAN

1. 实验目的

通过本实验可以掌握：
① PVLAN（私有 VLAN）的 VLAN 类型和端口类型。
② PVLAN 的配置。

2. 实验拓扑

私有 VLAN 配置实验拓扑如图 2-19 所示。

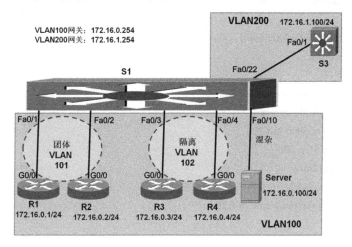

图 2-19 私有 VLAN 配置实验拓扑

3. 实验步骤

本实验把 R1～R4 以及 S3 作为用于测试的计算机使用。

（1）配置 R1～R4 以及 S3（供测试用）

```
R1(config)#no ip routing
R1(config)#ip default-gateway 172.16.0.254
//关闭路由器的路由功能，路由器相当于一台主机，因此需要配置网关
R1(config)#interface  GigabitEthernet0/0
R1(config-if)#no shutdown
R1(config-if)#ip address 172.16.0.1 255.255.255.0


R2(config)#no ip routing
R2(config)#ip default-gateway 172.16.0.254
R2(config)#interface GigabitEthernet0/0
R2(config-if)#no shutdown
```

R2(config-if)#**ip address 172.16.0.2 255.255.255.0**

R3(config)#**no ip routing**
R3(config)#**ip default-gateway 172.16.0.254**
R3(config)#**interface GigabitEthernet0/0**
R3(config-if)#**no shutdown**
R3(config-if)#**ip address 172.16.0.3 255.255.255.0**

R4(config)#**no ip routing**
R4(config)#**ip default-gateway 172.16.0.254**
R4(config)#**interface GigabitEthernet0/0**
R4(config-if)#**no shutdown**
R4(config-if)#**ip address 172.16.0.4 255.255.255.0**

S3(config)#**interface FastEthernet0/2**
S3(config-if)#**shutdown**
//关闭不必要的端口
S3(config)#**interface FastEthernet0/1**
S3(config-if)#**switch mode access**
S3(config-if)#**switch access vlan 200**
S3(config)#**interface vlan 200**
S3(config-if)#**no shutdown**
S3(config-if)#**ip address 172.16.1.100 255.255.255.0**
S3(config)#**ip default-gateway 172.16.1.254**

（2）配置 PVLAN

S1(config)#**vtp mode transparent**
//只有 VTP 模式为透明模式，才能配置 PVLAN
S1(config)#**vlan 101**
S1(config-vlan)#**private-vlan community**
//创建辅助 VLAN101，该 VLAN 是团体 VLAN
S1(config)#**vlan 102**
S1(config-vlan)#**private-vlan isolated**
//创建辅助 VLAN102，该 VLAN 是隔离 VLAN
S1(config)#**vlan 100**
S1(config-vlan)#**name VLAN-100**
S1(config-vlan)#**private-vlan primary**
S1(config-vlan)#**private-vlan association 101-102**
//创建主 VLAN100，把它和辅助 VLAN101 和 VLAN102 关联
S1(config)#**vlan 200**
S1(config-vlan)#**name VLAN-200**
//创建普通 VLAN200
S1(config)#**interface FastEthernet0/1**
S1(config-if)#**switchport private-vlan host-association 100 101**

S1(config-if)#**switchport mode private-vlan host**
//把 Fa0/1 端口配置为团体端口，它在 VLAN101 中
S1(config)#**interface FastEthernet0/2**
S1(config-if)#**switchport private-vlan host-association 100 101**
S1(config-if)#**switchport mode private-vlan host**
//把 Fa0/2 端口配置为团体端口，它在 VLAN101 中
S1(config)#**interface FastEthernet0/3**
S1(config-if)#**switchport private-vlan host-association 100 102**
S1(config-if)#**switchport mode private-vlan host**
//把 Fa0/3 端口配置为团体端口，它在 VLAN102 中
S1(config)#**interface FastEthernet0/4**
S1(config-if)#**switchport private-vlan host-association 100 102**
S1(config-if)#**switchport mode private-vlan host**
//把 Fa0/4 端口配置为团体端口，它在 VLAN102 中

S1(config)#**interface FastEthernet0/10**
S1(config-if)#**switchport mode private-vlan promiscuous**
S1(config-if)#**switchport private-vlan mapping 100 101,102**
//把 Fa0/10 端口配置为混杂端口，该端口可以和 VLAN101 和 VLAN102 中的计算机通信。该命令配置 Primary VLAN 和 Secondary VLAN 间的映射关系

S1(config)#**interface FastEthernet0/22**
S1(config-if)#**switchport access vlan 200**
S1(config-if)#**switchport mode access**
//把 Fa0/22 端口划分在普通 VLAN200 中，用于接入 S3

S1(config)#**ip routing**
S1(config)#**interface Vlan200**
S1(config-if)#**no shutdown**
S1(config-if)#**ip address 172.16.1.254 255.255.255.0**
S1(config)#**interface Vlan100**
S1(config-if)#**no shutdown**
S1(config-if)#**ip address 172.16.0.254 255.255.255.0**
//配置三层交换，S1 承担 VLAN100 和 VLAN200 间的路由功能，可参见后面的章节
S1(config-if)#**private-vlan mapping 101-102**
//将辅助 VLAN 映射到三层端口，允许 PVLAN 入口流量的三层交换

S1#**show interfaces f0/1 switchport**
Name: Fa0/1
Switchport: Enabled
Administrative Mode: private-vlan host
Operational Mode: private-vlan host
Administrative Trunking Encapsulation: negotiate
Operational Trunking Encapsulation: native

```
Negotiation of Trunking: Off
Access Mode VLAN: 1 (default)
Trunking Native Mode VLAN: 1 (default)
Administrative Native VLAN tagging: enabled
Voice VLAN: none
Administrative private-vlan host-association: 100 (VLAN-100) 101 (VLAN0101)
Administrative private-vlan mapping: none
Administrative private-vlan trunk native VLAN: none
Administrative private-vlan trunk Native VLAN tagging: enabled
Administrative private-vlan trunk encapsulation: dot1q
Administrative private-vlan trunk normal VLANs: none
Administrative private-vlan trunk private VLANs: none
Operational private-vlan:
    100 (VLAN-100) 101 (VLAN0101)
Trunking VLANs Enabled: ALL
Pruning VLANs Enabled: 2-1001
Capture Mode Disabled
Capture VLANs Allowed: ALL
//以上显示端口的配置情况
```

4. 实验调试

（1）团体 VLAN 内的计算机的连通性测试

```
R1#ping 172.16.0.2
Type escape sequence to abort.
Sending 5, 100-byte ICMP Echos to 172.16.0.2, timeout is 2 seconds:
!!!!!
Success rate is 80 percent (4/5), round-trip min/avg/max = 1/1/4 ms
```

R1 和 R2 在同一团体 VLAN 中，是可以通信的。

（2）隔离 VLAN 内的计算机的连通性测试

```
R3#ping 172.16.0.4
Type escape sequence to abort.
Sending 5, 100-byte ICMP Echos to 172.16.0.4, timeout is 2 seconds:
.....
Success rate is 0 percent (0/5)
```

R3 和 R4 在同一隔离 VLAN 中，是不可以通信的。

（3）和混杂端口上的计算机连的通性测试

```
R1#ping 172.16.0.100
Type escape sequence to abort.
Sending 5, 100-byte ICMP Echos to 172.16.0.100, timeout is 2 seconds:
!!!!!
Success rate is 80 percent (4/5), round-trip min/avg/max = 1/1/4 ms
```

```
R3#ping 172.16.0.100
Type escape sequence to abort.
Sending 5, 100-byte ICMP Echos to 172.16.0.100, timeout is 2 seconds:
!!!!!
Success rate is 80 percent (4/5), round-trip min/avg/max = 1/1/4 ms
```
R1 和 R3 都能和混杂端口上的计算机进行通信。

（4）团体 VLAN 中的计算机和隔离 VLAN 中的计算机间连通性测试

```
R1#ping 172.16.0.3
Type escape sequence to abort.
Sending 5, 100-byte ICMP Echos to 172.16.0.3, timeout is 2 seconds:
.....
Success rate is 0 percent (0/5)
```
R1 和 R3 不能进行通信。

（5）和其他普通 VLAN 中的计算机的连通性测试

```
R1#ping 172.16.1.100
Type escape sequence to abort.
Sending 5, 100-byte ICMP Echos to 172.16.1.100, timeout is 2 seconds:
!!!!!
Success rate is 80 percent (4/5), round-trip min/avg/max = 1/1/4 ms
```

```
R3#ping 172.16.1.100
Type escape sequence to abort.
Sending 5, 100-byte ICMP Echos to 172.16.1.100, timeout is 2 seconds:
!!!!!
Success rate is 80 percent (4/5), round-trip min/avg/max = 1/1/4 ms
```
团体 VLAN 中的 R1 和隔离 VLAN 中的 R3 都能和其他普通 VLAN 中的计算机进行通信。

2.10 本 章 小 结

本章首先介绍了交换机上的 VLAN 创建以及如何把端口划分到指定的 VLAN 中。在交换机之间级联的链路应该配置为 Trunk，Trunk 是否形成可以通过 DTP 协商，Trunk 有两种封装方式。VTP 可以让我们集中化管理 VLAN 的信息，VTP 有 3 种模式，不同模式可以完成不同的功能。EtherChannel 技术可以把多条链路捆绑起来形成大带宽的逻辑链路，EtherChannel 是否形成也可以用协议自动协商。最后还介绍了思科特有的私有 VLAN 技术，该技术有点复杂，但能够在一个 VLAN 中实现端口之间的隔离。

第 3 章　STP

为了减少网络的故障时间,我们经常会采用冗余拓扑。STP 可以让具有冗余结构的网络在发生故障时自动调整网络的数据转发路径。STP 有很多协议,早期的 IEEE 802.1d 重新收敛时间较长,通常需要 30~50 秒,为了减少这个时间,Cisco 公司引入了一些补充技术,例如,Uplinkfast、Backbonefast 和 Portfast 等。RSTP（IEEE 802.1w）则在协议上对 STP 进行根本的改进形成新的协议,从而减少收敛时间。STP 还有许多改进,例如,PVST+和 MSTP（IEEE 802.1s）协议,以及各种安全措施,本章将先介绍这些技术的基本原理,此后介绍常用的配置。

3.1　STP 协议概述

3.1.1　STP（IEEE 802.1d）简介

为了增加局域网的冗余性,我们常常会在网络中引入冗余链路,然而这样却会引起交换环路。交换环路会带来 3 个问题:广播风暴、同一帧的多个拷贝和交换机 CAM 表不稳定。STP（Spanning Tree Protocol）可以解决交换环路带来的这些问题,STP 的基本思路是阻断一些交换机端口,构建一棵没有环路的转发树。STP 利用 BPDU（Bridge Protocol Data Unit）和其他交换机进行协商,从而确定哪台交换机该阻断哪个端口。在 BPDU 中有几个关键的字段,例如,根桥 ID、根路径开销和端口 ID 等,BPDU 格式如图 3-1 所示。

2字节	协议ID	该值总为0
1字节	版本	STP 的版本（为 IEEE802.1d 时值为0）
1字节	BPDU类型	BPDU类型（配置BPDU=0, TCN BPDU=80）
1字节	标志	LSB（最低有效位）= TCN标志 MSB（最高有效位）= TCA标志
8字节	根ID	根桥的ID
4字节	根路径开销	到达根桥的STP开销
8字节	网桥ID	发送BPDU的网桥ID
2字节	端口ID	发送BPDU的网桥I端口ID
2字节	消息老化时间	根桥发送BPDU后的秒数,每经过一个网桥都会递减1,本质上它是到达根桥的跳计数
2字节	最大老化时间	根桥不可用之前保留根桥ID的最大时间
2字节	Hello时间	根桥连续发送BPDU的时间间隔
2字节	转发延迟	网桥在监听、学习状态所停留的时间

图 3-1　BPDU 格式

第 3 章 STP

为了在网络中形成一个没有环路的拓扑，网络中的交换机要完成以下 3 个步骤：① 选举根桥；② 选举根口；③ 选举指定口。在这些步骤中，哪个交换机能获胜将取决于以下因素（按顺序进行）：① 最小的根桥 ID；② 最小的根路径代价；③ 最小发送者桥 ID；④ 最小发送者端口 ID。

当网络的拓扑发生变化时，网络会从一个状态向另一个状态过渡，重新打开或阻断某些端口。交换机的端口要经过几种状态：禁用（Disable）、阻塞（Blocking）、监听状态（Listening）和学习状态（Learning）、最后是转发状态（Forwarding）。表 3-1 是 STP 各端口状态总结。

表 3-1 STP 各端口状态总结

状 态	阻 塞	监 听	学 习	转 发	禁 用
接收并处理 BPDU	能	能	能	能	不能
接收端口上收到的数据帧	不能	不能	不能	能	不能
发送其他端口交换过来的数据帧	不能	不能	不能	能	不能
学习 MAC 地址	不能	不能	能	能	不能

端口处于各种端口状态的时间长短取决于 BPDU 计时器。只有角色是根桥的交换机可以发送信息来调整计时器。以下计时器决定了 STP 的性能和状态转换：Hello 时间、转发延迟和最大老化时间。如图 3-2 所示是 STP 各个状态的转换过程。

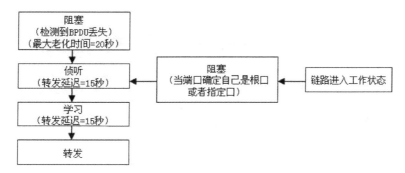

图 3-2 STP 各种状态的转换过程

如果检测到更改信息，交换机会通知生成树的根桥，然后根桥将该信息广播到整个网络。

3.1.2 STP 的加强

STP（IEEE 802.1d）的收敛时间通常需要 30～50 秒。为了减少收敛时间，有一些改善措施。Portfast 特性使得以太网端口一旦有设备接入，就立即进入转发状态，如果端口上连接的只是计算机或者其他不运行 STP 的设备，这是非常合适的。设置了 Portfast 后，端口 Up 或者 Down，交换机将不再发送 TCN 消息。

Uplinkfast 经常用在接入层交换机上，当它连接的主干交换机上的主链路出现故障时，能在 2～3 秒内切换到备份链路上，而不需要经过 30 秒或者 50 秒。Uplinkfast 只需要在接入层交换机上配置即可。Uplinkfast 可能导致严重不稳定，不应该在核心层或者分布层交换机上配置。

Backbonefast 主要用在主干交换机之间，当主干交换机之间的链路发生故障时，可以在 20～30 秒内就切换到备份链路上。Backbonefast 需要在全部交换机上配置。

3.1.3 PVST+简介

如图 3-3 所示，网络中有 2 个 VLAN，如果没有 PVST，两个 VLAN 的数据可能将全部在 Fa0/13 端口上的链路上传输，而另一链路空闲。采用 PVST（Per VLAN STP）后，PVST 会为每个 VLAN 构建一棵 STP 树。如图 3-4 所示，这样的好处是可以独立地为每个 VLAN 控制哪些端口要转发数据，从而实现负载平衡。图 3-4 中 VLAN1 的数据在 Fa0/13 端口上的链路上传输，VLAN2 的数据在 Fa0/14 端口上的链路上传输，大大提高了链路的利用率。PVST 的缺点是如果 VLAN 数量很多，由于要维护多个 STP 树，会给交换机带来一定的负担。由于 Cisco 的 PVST 实际上还集成了 STP 的加强功能，所以 Cisco 把它的 PVST 称为 PVST+，Cisco 交换机默认的模式就是 PVST+。

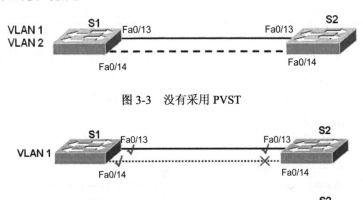

图 3-3 没有采用 PVST

图 3-4 采用 PVST 进行负载平衡

需要仔细规划每个 VLAN 的 STP 树，否则无法实现负载平衡。

3.1.4 RSTP（IEEE 802.1w）简介

RSTP（IEEE 802.1w）实际上是把减少 STP 收敛时间的一些加强措施融合在 STP 协议中形成新的协议。RSTP 能够达到相当快的收敛速度，有时甚至只需几百毫秒。RSTP 的特征如下：
① 集成了 IEEE 802.1d 的很多增强技术，这些增强功能不需要额外配置。
② RSTP 使用与 IEEE 802.1d 相同的 BPDU 格式。
③ RSTP 能够主动确认端口是否能安全转换到转发状态，而不需要依靠任何计时器来作出判断。

RSTP（IEEE 802.1w）使用第 2 版、第 2 类 BPDU，能够与 IEEE 802.1d 兼容。RSTP 发送 BPDU 以及填充标志字节的方式与 IEEE 802.1d 略有差异，RSTP BPDU 格式如图 3-5 所示。

2字节	协议ID=0x0000
1字节	协议版本ID=x02
1字节	BPDU类型=0x02
1字节	标志
8字节	根ID
4字节	根路径开销
8字节	网桥ID
2字节	端口ID
2字节	消息老化时间
2字节	最大老化时间
2字节	Hello时间
2字节	转发延迟

位	含义	
0	拓扑更改：TCN标志	
1	提议	
2~3	00	未知端口
	01	替代/备份端口
	10	根端口
	11	指定端口
4	学习	
5	转发	
6	同意	
7	拓扑更改确认：TCA标志	

图 3-5 RSTP BPDU 格式

① 如果连续 3 段 Hello 时间（默认值为 3×2= 6 秒）内没有收到 Hello 消息，或者当最大老化时间计时器过期时，协议信息可立即过期。与 STP 类似，RSTP 网桥会在每个 Hello 时间段（默认值为 2 秒）内发送包含其当前信息的 BPDU。然而 RSTP 网桥不是转发 BPDU，即使 RSTP 网桥没有从根桥收到任何 BPDU，RSTP 网桥也会可以产生 BPDU。

② 由于 BPDU 被用作保持活动的机制，连续 3 次未收到 BPDU 就表示网桥与其相邻的根桥或指定网桥失去连接。信息快速老化意味着故障能够被快速检测到。

RSTP 边缘端口是指永远不会用于连接到其他交换机设备的交换机端口。当启用时，此类端口会立即转换到转发状态，配置了"spanning-tree portfast"命令的端口就是边缘端口。非边缘端口分类为两种链路类型：点对点和共享。链路类型通常是自动确定的（全双工链路就是点对点类型，半双工就是共享类型），但可以使用配置命令进行指定。

RSTP 端口只有 3 种状态：丢弃、学习和转发。

① 丢弃：稳定的活动拓扑以及拓扑同步和更改期间都会出现此状态。丢弃状态禁止转发数据帧，因而可以断开第二层环路。

② 学习：稳定的活动拓扑以及拓扑同步和更改期间都会出现此状态。学习状态会接收数据帧来填充 MAC 表，以限制未知单播帧泛洪。

③ 转发：仅在稳定的活动拓扑中出现此状态。转发状态的交换机端口决定了拓扑。发生拓扑变化后，或在同步期间，只有当建议和同意过程完成后才会转发数据帧。

RSTP 端口除了可以是根口和指定口，对于非指定口还进一步分为替代（Alternate）端口和备份（Backup）端口。Alternate 端口由于收到其他网桥更优的 BPDU 而被阻塞，Backup 端口由于收到自己发出的更优的 BPDU 而被阻塞。如图 3-6 所示，S3 的 Fa0/1 是该网段的指定口，S2 将从 Fa0/1 接收 S3 发送的更优的 BPDU，所以为 Alternate 端口；S3 的 Fa0/2 将会接收到 S3 自己发送的更优的 BPDU，所以为 Backup 端口。当 S2 的根口（RP）发生故障时，S2 的替代（AP）端口将立即进入转发状态；而当 S3 的指定端口故障时，S2 的备份端口（BP）将立即进入转发状态，从而大大减少收敛时间。

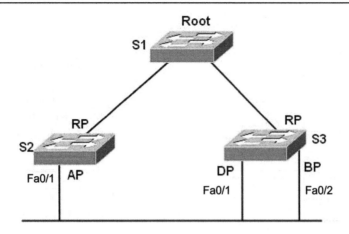

图 3-6 替代端口（Alternate）和备份端口（Backup）示意图

RSTP 使用提议／同意握手机制来完成端口的快速收敛。如图 3-7 所示，假设 SW1 有一条新的链路连接到根桥，当链路处于 Up 状态时，根桥的 p0 口和 SW1 的 p1 口同时进入指定阻断状态，而且 p0 和 p1 同时发布带有提议标志位的 RSTP BPDU，同时 p1 成为新的根端口。SW1 开始同步新的消息给其他的端口，p2 为替换端口，在同步过程中保持不变。p3 为指定端口，在过程同步中必须阻断。p4 为边缘端口，在同步过程中保持不变。SW1 通过新的根端口 p1 给根桥发送一个提议 BPDU 同意消息，即将同意标志位设为同意，p0 和 p1 握手成功，p0 和 p1 直接进入转发状态。这时 p3 端口为指定端口，还处于阻断状态。同样，按照 p0 和 p1 的提议／同意握手机制，p3 快速进入转发状态。提议／同意握手机制收敛很快，状态转变中无须依赖任何定时器。

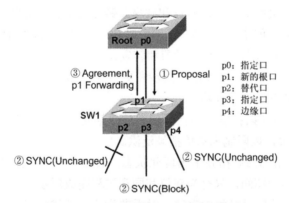

图 3-7 RSTP 使用提议／同意握手机制

3.1.5 MSTP（IEEE 802.1s）简介

在 PVST 中，交换机为每个 VLAN 都构建一棵 STP 树，不仅会给 CPU 带来很大负载（特别是低端的交换），也会占用大量的带宽。MSTP（Multiple Spanning Tree Protocol）则是把多个 VLAN 映射到一个 STP 实例上，从而减少了 STP 实例。MSTP 可以与 STP 和 PVST 配合使用。对于运行 STP 和 PVST 的交换机来说，一个 MSTP 域看起来就像一台交换机。

1. 实例和域

在 MSTP 中引入了"实例"（Instance）和"域"（Region）的概念。所谓"实例"就是多个 VLAN 的一个集合，这种通过多个 VLAN 捆绑到一个实例中的方法可以节省通信开销和资源占用率。MSTP 各个实例拓扑的计算是独立的，控制这些实例就可以实现负载均衡。所谓"域"由域名（Configuration Name）、修订级别（Revision Level）、格式选择器（Configuration Identifier Format Selector）、VLAN 与实例的映射关系（Mapping of VIDs to Spanning Trees）组成，其中域名、格式选择器和修订级别在 BPDU 报文中都有相关字段，而 VLAN 与实例的映射关系在 BPDU 报文中表示为摘要信息（Configuration Digest），该摘要是根据映射关系计算得到的一个 16 字节签名。只有上述四者都一样，相互连接的交换机才认为在同一个域内。MSTP 区域如图 3-8 所示，每个域内所有交换机都有相同的 MSTP 域配置。默认时，所有的 VLAN 都映射到实例 0 上。

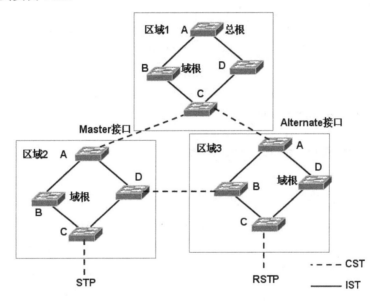

图 3-8　MSTP 区域

MSTP 的实例 0 具有特殊的作用，称为 CIST（Common Internal and Spanning Tree），即公共和内部生成树，其他的实例称为 MSTI（Multiple Spanning Tree Instance），即多生成树实例。CIST 是每个 MST 区域中的 IST 实例、互连 MST 区域的 CST 实例和 IEEE 802.1d 网桥的集合，是为了保证所有桥接的局域网是简单的和全连接的。CST（Common Spanning Tree）是 STP/RSTP 也是 MSTP 计算出的用于连接 MSTP 域的单生成树。IST（Internal Spanning Tree）在一个给定的 MSTP 域内由 CIST 提供连通性。如图 3-8 所示，如果把每个 MSTP 区域看作一个"交换机"，CST 就是这些"交换机"通过 STP/RSTP 或者 MSTP 协议计算生成的一棵生成树。IST 是 CIST 在 MSTP 区域内的片段，是一个特殊的多生成树实例。

2. 总根和域根

如图 3-8 所示，与 STP 和 RSTP 相比，MSTP 中引入了总根和域根的概念。总根是一个全局概念，对于所有互连的运行 STP/RSTP/MSTP 的交换机只能有一个总根，也即是 CIST

的根;而域根是一个局部概念,是相对于某个域的某个实例而言的。如图 3-8 所示,所有相连的设备,总根只有一个,而每个域所包含的域根数目与实例个数相关。

3. 外部路径开销和内部路径开销

与 STP 和 RSTP 相比,MSTP 中引入了外部路径开销和内部路径开销的概念。外部路径开销是相对于 CIST 而言的,同一个域内外部路径开销是相同的;内部路径开销是域内相对于某个实例而言的,同一端口对于不同实例对应不同的内部路径开销。

4. 边缘端口、Master 端口和 Alternate 端口

与 STP 和 RSTP 相比,MSTP 中引入了域边缘端口和 Master 端口的概念。域边缘端口是连接不同 MSTP 域、MSTP 域和运行 STP 的区域及 MSTP 域和运行 RSTP 的区域的端口,位于 MSTP 域的边缘。在某个不包含总根的域中,Master 端口是所有边界端口中到达总根开销最小的端口,也就是连接 MSTP 域到总根的端口,位于整个域到总根的最短路径上。Alternate 端口是 Master 端口的替代端口,如果 Master 端口被阻塞,Alternate 端口将成为新的 Master 端口。如图 3-8 所示,域根在区域 1 中,其中设备 C 与区域 2 和区域 3 相连的端口是域边界端口,而在区域 2 中,设备 A 与区域 1 相连的端口是 Master 端口;在区域 3 中,设备 A 与区域 1 相连的端口是 Alternate 端口。

3.1.6 不同 STP 协议的兼容性

1. IEEE 802.1d 与 IEEE 802.1w 的兼容性

STP(IEEE 802.1d)无法知道 RSTP(IEEE 802.1w)中 BPDU 带有的版本号为 2,其本身的版本号为 0,但是 RSTP 可以识别版本号为 0 的 STP,一旦 RSTP 的端口连接的是 STP 的设备,该端口将使用 STP 的 BPDU 和 TCN 来运行,以保证 RSTP 和 STP 的互操作性。但是将 RSTP 和 STP 结合工作,RSTP 将失去其固有的快速收敛的能力。

2. IEEE 802.1s 与 IEEE 802.1d 和 IEEE 802.1w 的兼容性

IEEE 802.1s 与 IEEE 802.1d 和 IEEE 802.1w 兼容。

要说明的是不同 STP 的协议虽然可以兼容,然而不同 STP 结合工作会导致问题复杂化,强烈建议不要在一个网络中使用不同的 STP 协议。

3.1.7 STP 防护

1. STP 树防护

STP 协议并没有什么措施对交换机的身份进行认证。在稳定的网络中,如果接入非法交换机将可能给网络中的 STP 树带来灾难性的破坏。有一些简单的措施来保护网络,虽然这些措施显得有点软弱无力。Root Guard 特性将使得交换机的端口拒绝接收比原有根桥优先级更高的 BPDU。而 BPDU Guard 主要和 Portfast 特性配合使用,Portfast 使得端口一有设备接入

就立即进入转发状态，然而万一这个端口接入的是交换机很可能造成交换环路。BPDU Guard 可以使 Portfast 端口一旦接收到 BPDU，就关闭该端口。

BPDU Filter 也是和 Portfast 配合使用的，配置了 Portfast 的端口通常接计算机，在这些端口配置了 BPDU Filter 后，将不再发送和接收 BPDU。BPDU Filter 可以在全局模式下配置，也可以在端口下配置。BPDU Filter 状态启用与否和以下因素相关：端口的 BPDU Filter 配置、全局的 BPDU Filter 配置和端口的 Portfast 配置，BPDU Filter 状态结果如表 3-2 所示。

表 3-2 BPDU Filter 状态结果

端口的 BPDU Filter 配置	全局配置的 BPDU Filter	Portfast 状态	最终的 BPDU Filter 状态
启用	—	—	启用
禁用	—	—	禁用
默认（即无配置）	默认（即无配置）	—	禁用
默认（即无配置）	启用	禁用	禁用
默认（即无配置）	启用	启用	启用，但一旦收到 BPDU，Portfast 和 BPDU Filter 功能将丧失

2. 环路保护

光纤或者 UTP 双绞线中有可能出现单向传输的故障。如图 3-9 所示，交换机 S3 的 Fa0/2 端口原处于替代端口状态，如果 S2 和 S3 之间的链路出现单向故障（S2 不能发送数据包到 S3，但是 S3 可以发送数据包到 S2），则 S3 的 Fa0/2 端口将最终进入转发状态，从而导致交换环路。可以采用 Guard Loop 和 UDLD 技术避免环路。

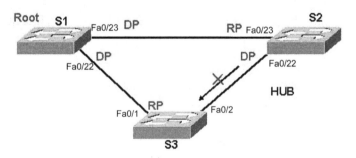

图 3-9 单向链路引起交换环路

Guard Loop 应该在所有非指定端口上配置，则这些端口在经过最大寿命计时器之后（默认值为 20 秒），将过渡到 Broken 状态，而不是进入转发状态，从而避免交换环路。Guard Loop 实际上阻止了非指定端口变为指定端口，笔者认为该技术阻止了 STP 实现冗余的功能，万一不是出现单向故障而是真正的线路故障，STP 将无法发挥作用。

UDLD 技术也可以避免链路单向故障而引起的交换环路，但是 UDLD 并不是 STP 的一部分。当 UDLD 检测到单向故障后，可以把端口关闭（Err-disable），这样端口就不会收发数据了，环路自然就没有了。UDLD 在每个活动的端口上周期性地发送 Probe/Echo 包，以维护邻居缓存的完整性。一端收到邻居发送的 Hello 信息后会将它缓存到内存中，并保存一个通过 Hold-time 定义的时间间隔。如果 Cache 超时，相应的 Cache 被清空，如果在 Hold-time 定义的时间间隔内收到邻居新的 Hello 信息，则用新的信息替代老的，并将计时器清零。为了维护 UDLD Cache 的完整性，一旦一个启用了 UDLD 的端口被禁用，或该端口上的设备被重启,该端口存在的所有缓存均会被清除,UDLD 传送至少一个信息让邻居清空响应的缓存条目。在启用 UDLD 的情况下,交换机将定期地向邻居发送 UDLD 协议数据包,并且期望在预定计时器到期之前接收到回应数据包。如果计时器到期,那么交换机将确定该链路是单向链路,并且关闭该端口。

Guard Loop 和 UDLD 比较如表 3-3 所示。

表 3-3 Guard Loop 和 UDLD 比较

	Guard Loop	UDLD
配置	按端口	按端口
活动粒度	按 VLAN	按端口
自动恢复	是	是(要设置 Err-disable Recovery)
防止单向链路导致交换环路	是	是
防止软件问题导致指定端口不发送 BPDU	是	否
防止错误配线	否	是

3.1.8 FlexLink

FlexLink 能够提供二层永续性,是 STP 的替代方案,可以让用户在关闭 STP 的情况下仍然实现冗余功能。一般在接入交换机和分布交换机之间运行,但是是配置在接入层交换机上的,它的收敛时间优于生成树协议 / 快速生成树协议 / IEEE 802.1w,收敛时间低于 100 ms。如图 3-10(交换机 S3)所示,FlexLink 定义了一对主 / 备用端口,只有一个端口处于启用状态。如果主用端口 Fa0/1 发生故障,备用端口 Fa0/2 开始转发。这两个端口的任一个可以是二层的接入端口和 EtherChannel 端口,三层端口不支持 FlexLink。在 FlexLink 端口上是禁用 STP 的,它们不参与 STP。FlexLink 原理如图 3-10 所示。

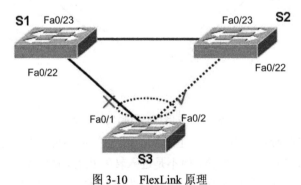

图 3-10 FlexLink 原理

3.2 实验 1:STP 和 PVST 配置

1. 实验目的

通过本实验可以掌握:
① STP 的工作原理。
② STP 树的控制方法。

③ 利用 PVST 进行负载平衡的方法。

2. 实验拓扑

在图 3-11 中，S1 和 S2 模拟核心层交换机，而 S3 为接入层交换机，本实验不使用 R1，3 台交换机实际上是三层交换机，这里我们并不利用其三层功能，所以采用二层交换机的图标。Cisco 交换机默认是运行 PVST+的，因此每个 VLAN 有一棵 STP 树。

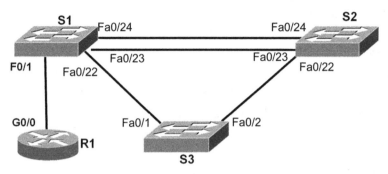

图 3-11　STP 和 PVST 实验拓扑

3. 实验步骤

我们要在网络中配置 2 个 VLAN，不同 VLAN 的 STP 具有不同的根桥，实现负载平衡。

（1）利用 VTP 在交换机上创建 VLAN2，S1 和 S2 之间的链路配置 Trunk

```
S1(config)#interface fastEthernet0/23
S1(config-if)#switchport trunk encapsulation dot1q
S1(config-if)#switchport mode trunk
S1(config)#interface fastEthernet0/24
S1(config-if)#switchport trunk encapsulation dot1q
S1(config-if)#switchport mode trunk
S1(config-if)#shutdown    //先关闭该端口，暂时不使用该链路
S1(config)#interface fastEthernet0/22
S1(config-if)#switchport trunk encapsulation dot1q
S1(config-if)#switchport mode trunk

S2(config)#interface fastEthernet0/23
S2(config-if)#switchport trunk encapsulation dot1q
S2(config-if)#switchport mode trunk
S2(config)#interface fastEthernet0/24
S2(config-if)#switchport trunk encapsulation dot1q
S2(config-if)#switchport mode trunk
S2(config-if)#shutdown    //先关闭该接口，暂时不使用该链路
S2(config)#interface fastEthernet0/22
S2(config-if)#switchport trunk encapsulation dot1q
S2(config-if)#switchport mode trunk
```

```
S3(config)#interface fastEthernet0/1
S3(config-if)#switchport trunk encapsulation dot1q
S3(config-if)#switchport mode trunk
S3(config)#interface fastEthernet0/2
S3(config-if)#switchport trunk encapsulation dot1q
S3(config-if)#switchport mode trunk
//请使用"show interfaces trunk"命令确认各链路的 Trunk 是否成功
S1(config)#vtp domain VTP-TEST
Changing VTP domain name from NULL to VTP-TEST
S1(config)#vlan 2
```

在 S1 上配置 VTP 的域名，创建 VLAN 2。由于默认时 S2 和 S3 的 VTP 域名为空，它们将自动学习到 S1 的 VTP 域名，同时 S2 和 S3 也将自动学习到 VLAN 2，请在 S2 和 S3 交换机上使用"show vlan"命令确认 VLAN 是否创建成功。

（2）检查默认的 STP 树

```
S1#show spanning-tree
VLAN0001
  Spanning tree enabled protocol ieee    //表明运行的 STP 协议是 IEEE 802.1d
  Root ID    Priority    32769
//根桥的优先级，默认值为 32768，之所以加 1，是因为这是 VLAN1 的 STP
             Address     0023.ac7d.6c80   //根桥的 MAC 地址
             This bridge is the root      //根桥就是本交换机
             Hello Time   2 sec   Max Age 20 sec   Forward Delay 15 sec
//以上是各个计时器的值（Hello 时间、老化时间和转发时间）

  Bridge ID  Priority    32769   (priority 32768 sys-id-ext 1)
             Address     0023.ac7d.6c80
             Hello Time  2 sec   Max Age 20 sec   Forward Delay 15 sec
             Aging Time 300
//以上显示该交换机的桥 ID 等信息，由于本交换机就是根桥，所以这些信息和上一段落显示的信息是一样的

Interface          Role Sts Cost        Prio.Nbr Type
---------------- ---- --- ---------  -------- --------------------------------
Fa0/1              Desg FWD 19          128.3    P2p
Fa0/3              Desg FWD 19          128.5    P2p
Fa0/4              Desg FWD 19          128.6    P2p
Fa0/10             Desg FWD 19          128.12   P2p
Fa0/22             Desg FWD 19          128.24   P2p    //该端口是指定口
Fa0/23             Desg FWD 19          128.25   P2p    //该端口是指定口
```

以上显示该交换机各个端口的状态，Fa0/22 和 Fa0/23 为转发状态。"Role"列是端口的角色，Desg 是指定口，Altn 是 Alternate 指定口，Root 是根口；"Sts"列是端口的状态，FWD 表示在转发，BLK 表示在阻断，LIS 表示在监听，LRN 表示在学习 MAC 地址；"Cost"列是

端口的 Cost 值;"Prio.Nbr"列是端口的优先级;"Type"列是端口的类型,P2p 表示是点对点类型,Shr 表示是共享类型。

```
VLAN0002
  Spanning tree enabled protocol ieee
  Root ID    Priority      32770
             Address       0023.ac7d.6c80
             This bridge is the root
             Hello Time   2 sec   Max Age 20 sec   Forward Delay 15 sec

  Bridge ID  Priority      32770  (priority 32768 sys-id-ext 2)
             Address       0023.ac7d.6c80
             Hello Time   2 sec   Max Age 20 sec   Forward Delay 15 sec
             Aging Time   15

Interface         Role Sts Cost      Prio.Nbr Type
----------------- ---- --- --------- --------------------------------
Fa0/22            Desg FWD 19        128.24   P2p
Fa0/23            Desg FWD 19        128.25   P2p
```

以上是 VLAN2 的 STP 树情况,VLAN2 的 STP 树和 VLAN1 的类似。默认时,Cisco 交换机会为每个 VLAN 都生成一个单独的 STP 树,称为 PVST(Per VLAN Spanning Tree)。

提示

我们现在并没有采用 RSTP,但是端口上显示的角色和类型却是 RSTP 的信息,这不影响 STP。

```
S2#show spanning-tree
VLAN0001
  Spanning tree enabled protocol ieee
  Root ID    Priority      32769
             Address       0023.ac7d.6c80
             Cost          19
             Port          25 (FastEthernet0/23)
             Hello Time   2 sec   Max Age 20 sec   Forward Delay 15 sec
//以上显示根交换机的信息(交换机 S1)
  Bridge ID  Priority      32769  (priority 32768 sys-id-ext 1)
             Address       0023.ac9d.f000
             Hello Time  2 sec   Max Age 20 sec   Forward Delay 15 sec
             Aging Time  300

Interface         Role Sts Cost      Prio.Nbr Type
----------------- ---- --- --------- --------------------------------
Fa0/2             Desg FWD 19        128.4    P2p
Fa0/3             Desg FWD 19        128.5    P2p
Fa0/4             Desg FWD 19        128.6    P2p
Fa0/22            Altn BLK 19        128.24   P2p   //该端口被阻断
Fa0/23            Root FWD 19        128.25   P2p   //该端口是根口
```

```
VLAN0002
  Spanning tree enabled protocol ieee
  Root ID      Priority     32770
               Address      0023.ac7d.6c80
               Cost         19
               Port         25 (FastEthernet0/23)
               Hello Time   2 sec   Max Age 20 sec   Forward Delay 15 sec
  Bridge ID    Priority     32770   (priority 32768 sys-id-ext 2)
               Address      0023.ac9d.f000
               Hello Time   2 sec   Max Age 20 sec   Forward Delay 15 sec
               Aging Time  300

Interface           Role Sts Cost      Prio.Nbr Type
------------------- ---- --- --------- -------- --------------------------------
Fa0/22              Altn BLK 19        128.24   P2p    //该端口被阻断
Fa0/23              Root FWD 19        128.25   P2p    //该端口是根口

S3#show spanning-tree vlan 1
VLAN0001
  Spanning tree enabled protocol ieee
  Root ID      Priority     32769
               Address      0023.ac7d.6c80
               Cost         19
               Port         3 (FastEthernet0/1)
               Hello Time   2 sec   Max Age 20 sec   Forward Delay 15 sec
//以上显示根交换机的信息（交换机 S1）
  Bridge ID    Priority     32769   (priority 32768 sys-id-ext 1)
               Address      0023.ac7d.9c00
               Hello Time   2 sec   Max Age 20 sec   Forward Delay 15 sec
               Aging Time  300

Interface           Role Sts Cost      Prio.Nbr Type
------------------- ---- --- --------- -------- --------------------------------
Fa0/1               Root FWD 19        128.3    P2p    //该端口是根口
Fa0/2               Desg FWD 19        128.4    P2p    //该端口是指定口

VLAN0002
  Spanning tree enabled protocol ieee
  Root ID      Priority     32770
               Address      0023.ac7d.6c80
               Cost         19
               Port         3 (FastEthernet0/1)
               Hello Time   2 sec   Max Age 20 sec   Forward Delay 15 sec
  Bridge ID    Priority     32770   (priority 32768 sys-id-ext 2)
               Address      0023.ac7d.9c00
```

	Hello Time 2 sec	Max Age 20 sec	Forward Delay 15 sec	
	Aging Time 300			
Interface	Role Sts Cost	Prio.Nbr Type		
Fa0/1	Root FWD 19	128.3	P2p	//该接口是根口
Fa0/2	Desg FWD 19	128.4	P2p	//该接口是指定口

📖 技术要点

STP 的结果如图 3-12（只考察 VLAN1），需要仔细分析为什么 STP 会是目前这种情况。3 台交换机的默认优先级都是 32768，而 S1 的 MAC 地址较小，所以成为了根桥，则 S1 上的 Fa0/22 和 Fa0/23 是指定口，处于 Forword 状态。S2 有两个端口可以到达 S1，一个端口是 Fa0/22，到达 S1 的 Cost 为 19+19=38；另一个端口是 Fa0/23，到达 S1 的 Cost 为 19，因此 Fa0/23 是根口，处于 Forword 状态。同样，在 S3 上，Fa0/1 也是根口，处于 Forword 状态。在 S2 和 S3 之间的链路上，要选举出一个指定口。根据选举的要素，根桥的 ID 是一样的，不能决出胜负；到达根桥的 Cost 值也是一样的，都为 19，不能决出胜负；但是发送者桥 ID 不一样，S2 的 MAC 地址高，S3 的 MAC 地址低，S3 获胜，所以 S3 的 Fa0/2 是指定口，处于 Forward 状态，S2 的 Fa0/22 就处于 Block 状态了。

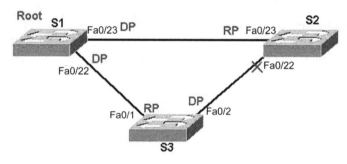

图 3-12　VLAN1 默认的 STP 树

（3）控制 S1 为 VLAN1 的根桥

当前的 S1 已经是根桥了，然而是因为其 MAC 地址低而获胜成为根桥。如果哪天交换机损坏进行更换，MAC 地址发生变化，并不能保证 S1 还是根桥。因此有必要通过优先级来确保它是根桥。

```
S1(config)#spanning-tree vlan 1 priority 4096
//对于 VLAN1 来说，把 S1 的优先级为 4096，而 S2 和 S3 保持默认值 32768，这样 S1 就成为了 VLAN1 的根桥
S1#show spanning-tree vlan 1
VLAN0001
  Spanning tree enabled protocol ieee
  Root ID    Priority    4097
             Address     0023.ac7d.6c80
             This bridge is the root
             Hello Time  2 sec  Max Age 20 sec  Forward Delay 15 sec
  Bridge ID  Priority    4097    (priority 4096 sys-id-ext 1)   //优先级被更改了
```

```
              Address     0023.ac7d.6c80
              Hello Time  2 sec   Max Age 20 sec   Forward Delay 15 sec
              Aging Time 300

Interface          Role Sts Cost      Prio.Nbr Type
------------------ ---- --- --------- --------------------------------
Fa0/1              Desg FWD 19        128.3    P2p
Fa0/3              Desg FWD 19        128.5    P2p
Fa0/4              Desg FWD 19        128.6    P2p
Fa0/10             Desg FWD 19        128.12   P2p
Fa0/22             Desg FWD 19        128.24   P2p
Fa0/23             Desg FWD 19        128.25   P2p
```

（4）控制指定口

从步骤（3）中可以看到，对于 VLAN1，S1 成为根桥，S1 的 Fa0/22 和 Fa0/23 处于转发状态；S2 的 Fa0/23 是根口，也处于转发状态；S3 的 Fa0/1 是根口，也处于转发状态；然而在 S2 和 S3 之间的链路上，却是低端交换机 S3 的 Fa0/2 在转发数据，原因在于 S2 和 S3 在竞争指定口时，由于 S3 的 MAC 地址较低而获胜了，这是不合理的。我们要控制指定口，可以通过改变优先级实现，操作如下：

S2(config)#spanning-tree vlan 1 root secondary

配置 S2 的优先级仅比根桥低一点，但比其他交换机高。对于 VLAN1 来说，S2 的优先级比 S1 低（值越大，优先级越低），不至于成为根桥，但是比 S3 高，所以在竞争指定口时会获胜。

```
S2#show spanning-tree vlan 1
VLAN0001
  Spanning tree enabled protocol ieee
  Root ID      Priority    4097
               Address     0023.ac7d.6c80
               Cost        19
               Port        25 (FastEthernet0/23)
               Hello Time  2 sec   Max Age 20 sec   Forward Delay 15 sec

  Bridge ID    Priority    28673   (priority 28672 sys-id-ext 1)   //修改后的优先级
               Address     0023.ac9d.f000
               Hello Time  2 sec   Max Age 20 sec   Forward Delay 15 sec
               Aging Time 15

Interface          Role Sts Cost      Prio.Nbr Type
------------------ ---- --- --------- --------------------------------
Fa0/2              Desg FWD 19        128.4    P2p
Fa0/3              Desg FWD 19        128.5    P2p
Fa0/4              Desg FWD 19        128.6    P2p
Fa0/22             Desg FWD 19        128.24   P2p    // S2 上的 Fa/22 处于转发状态
Fa0/23             Root FWD 19        128.25   P2p

S3#show spanning-tree vlan 1
```

第 3 章 STP

```
VLAN0001
  Spanning tree enabled protocol ieee
  Root ID      Priority     4097
               Address      0023.ac7d.6c80
               Cost         19
               Port         3 (FastEthernet0/1)
               Hello Time   2 sec  Max Age 20 sec   Forward Delay 15 sec
  Bridge ID    Priority     32769  (priority 32768 sys-id-ext 1)
               Address      0023.ac7d.9c00
               Hello Time   2 sec  Max Age 20 sec   Forward Delay 15 sec
               Aging Time 300
Interface           Role Sts Cost      Prio.Nbr Type
---------------- ---- --- --------- -------- --------------------------------
Fa0/1               Root FWD 19        128.3    P2p
Fa0/2               Altn BLK 19        128.4    P2p    // S3 上的 Fa0/2 处于阻断状态
```

提示

"spanning-tree vlan 1 root secondary" 和 "spanning-tree vlan 1 root primary" 命令实际上是宏命令，当执行该命令时，交换机会先取出当前根桥的优先级，然后把本交换机的优先级设置为比当前根桥优先级大或者小 4096 的值（值越大，优先级越低）。在使用"show running"命令时，在配置文件中是看不到该命令的。

（5）配置 PVST 使得 S2 是 VLAN 2 的根桥

```
S2(config)#spanning-tree vlan 2 priority 4096
S1(config)#spanning-tree vlan 2 root secondary
S3#show spanning-tree vlan 1
（省略）
Interface           Role Sts Cost      Prio.Nbr Type
---------------- ---- --- --------- -------- --------------------------------
Fa0/1               Root FWD 19        128.3    P2p
Fa0/2               Altn BLK 19        128.4    P2p
S3#show spanning-tree vlan 2
（省略）
Interface           Role Sts Cost      Prio.Nbr Type
---------------- ---- --- --------- -------- --------------------------------
Fa0/1               Altn BLK 19        128.3    P2p
Fa0/2               Root FWD 19        128.4    P2p
```

从以上可以看到，在 VLAN1 和 VLAN2 的 STP 树中，端口状态不一样。STP 树的结果如图 3-13 所示，可以看到用 PVST 实现了 VLAN1 和 VLAN2 的负载平衡。

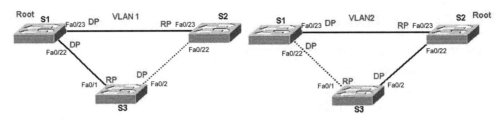

图 3-13　用 PVST 实现负载平衡

（6）启用 S1 和 S2 之间的 Fa0/24 链路

```
S1(config)#interface fastEthernet0/24
S1(config-if)#no shutdown
S1(config-if)#switchport trunk encapsulation dot1q
S1(config-if)#switchport mode trunk
S2(config)#interface fastEthernet0/24
S2(config-if)#no shutdown
S2(config-if)#switchport trunk encapsulation dot1q
S2(config-if)#switchport mode trunk
```

用"**show spanning-tree vlan**"命令，可以看到图 3-14 所示的 STP 树（树的一部分，我们现在只关心 S1 和 S2 之间的链路），如果 VLAN 1 和 VLAN 2 的数据要在 S1 和 S2 之间传输，将全部在 Fa0/23 之间的链路上传输。这样不是很合理，我们也要进行负载平衡。

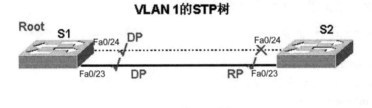

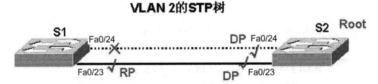

图 3-14　交换机 S1 和 S2 的部分 STP 树

有两种方法，第一种方法是修改 Cost 值，如下所示：

```
S2(config)#interface fastEthernet0/24
S2(config-if)#spanning-tree vlan 1 cost 18
```

在 S2 上修改 Fa0/24 的 Cost 值。由于 Fa0/23 上的 Cost 默认值为 19，因此 Fa0/24 端口成为了根口，而 Fa0/23 被阻断了。新的 STP 树（部分）如图 3-15 所示，VLAN2 的 STP 没有变化。S1 和 S2 之间的 VLAN1 数据通过 Fa0/24 链路传输；VLAN 2 数据通过 Fa0/23 链路传输。

图 3-15 新的 STP 树（部分）

第二种方法是修改端口的优先级，如下所示：

S2(config)#**interface fastEthernet0/24**
S2(config-if)#**no spanning-tree vlan 1 cost 18**   //恢复端口的 Cost 值
S1(config)#**interface fastEthernet0/24**
S1(config-if)#**spanning-tree vlan 1 port-priority 112**

在端口上配置优先级。要注意的是：需要在 S1 上配置，在 S2 的 Fa0/24 上配置是不起作用的。端口优先级默认值为 128，值越小，优先级越高，优先级要是 16 的倍数。STP 的结果和图 3-15 所示是一样的。

提示

在以上实验中我们用 PVST 来实现 S1 和 S2 之间两条链路的负载平衡。实际上把 Fa0/23 和 Fa0/24 捆绑形成 EtherChannel 应该会更好，EtherChannel 不仅有冗余功能，还能实现负载平衡，而且切换时间更短。关于 EtherChannel 请见前面的章节。

技术要点

在实际中，当要显式控制根桥和根口等信息时，不要利用交换机的 MAC 地址来控制，如果交换机更换或者网络中加入新的交换机，由于这些交换机的 MAC 地址并不能事先知道，可能导致 STP 和我们预想的不一致。另外，可以使用 "**no spanning-tree vlan XXX**" 命令关闭某个 VLAN 的 STP 树，然而请确保 VLAN 不存在交换环路。强烈建议不要做此动作，STP 对交换机带来的负载通常是可以忍受的。

4. 实验调试

（1）debug spanning-tree bpdu

该命令用来调试跟踪 STP 接收 BPDU 的情况。该命令只会显示交换机接收到的 BPDU。

S1#**debug spanning-tree bpdu**
    *Mar   1 01:28:22.690: STP: VLAN0002 rx BPDU: config protocol = ieee, packet from FastEthernet0/23, linktype SSTP, enctype 3, encsize 22
    *Mar   1 01:28:22.690: STP: enc 01 00 0C CC CC CD 00 23 AC 9D F0 19 00 32 AA AA 03 00 00 0C 01 0B
    *Mar   1 01:28:22.690: STP: Data 000000000010020023AC9DF0000000000010020023AC9DF00080190000140002000F00
    *Mar   1 01:28:22.690: STP: VLAN0002 Fa0/23:0000 00 00 00 10020023AC9DF000 00000000

10020023AC9DF000 8019 0000 1400 0200 0F00

　　*Mar　1 01:28:22.690: STP(2) port Fa0/23 supersedes 0

　　*Mar　1 01:28:22.690: STP: VLAN0002 rx BPDU: config protocol = ieee, packet from FastEthernet0/24, linktype SSTP , enctype 3, encsize 22

　　*Mar　1 01:28:22.690: STP: enc 01 00 0C CC CC CD 00 23 AC 9D F0 1A 00 32 AA AA 03 00 00 0C 01 0B

　　*Mar　1 01:28:22.690: STP: Data 000000000010020023AC9DF0000000000010020023AC9DF000801A0000140002000F00

　　*Mar　1 01:28:22.690: STP: VLAN0002 Fa0/24:0000 00 00 00 10020023AC9DF000 00000000 10020023AC9DF000 801A 0000 1400 0200 0F00

（2）debug spanning-tree events

该命令用来调试跟踪 STP 的事件。

S1#**debug spanning-tree events**
Spanning Tree event debugging is on　　//跟踪 STP 的事件
S1(config)#**interface fastEthernet0/22**
S1(config-if)#**shutdown**
S1(config-if)#**no shutdown**　　//把 Fa0/22 端口关闭后重新打开，观察 STP 的运行情况
*Mar　1 01:36:22.871: set portid: VLAN0001 Fa0/22: new port id 8018
*Mar　1 01:36:22.871: STP: VLAN0001 Fa0/22 -> **listening**　　//进入监听状态
*Mar　1 01:36:22.871: set portid: VLAN0002 Fa0/22: new port id 8018
*Mar　1 01:36:22.871: STP: VLAN0002 Fa0/22 -> listening
*Mar　1 01:36:24.557: STP: VLAN0001 Topology Change rcvd on Fa0/22
*Mar　1 01:36:37.878: STP: VLAN0001 Fa0/22 -> **learning**　　//进入学习状态
*Mar　1 01:36:37.878: STP: VLAN0002 Fa0/22 -> learning
*Mar　1 01:36:52.885: STP: VLAN0001 Fa0/22 -> **forwarding**　　//进入转发状态
*Mar　1 01:36:52.885: STP: VLAN0002 sent Topology Change Notice on Fa0/23
*Mar　1 01:36:52.885: STP: VLAN0002 Fa0/22 -> forwarding

（3）show spanning-tree vlan 1 detail

S1#**show spanning-tree vlan 1 detail**
　VLAN0001 is executing the ieee compatible Spanning Tree protocol
　　Bridge Identifier has priority 4096, sysid 1, address 0023.ac7d.6c80
　　Configured hello time 2, max age 20, forward delay 15　　//配置各个计时器
　　We are the root of the spanning tree
　　Topology change flag not set, detected flag not set
　　Number of topology changes 10 last change occurred 00:07:38 ago　　//拓扑发生变化次数
　　　　from FastEthernet0/22
　　Times:　hold 1, topology change 35, notification 2
　　　　　　hello 2, max age 20, forward delay 15
　　Timers: hello 1, topology change 0, notification 0, aging 300
　Port 3 (FastEthernet0/1) of VLAN0001 is designated forwarding
　　Port path cost 19, Port priority 128, Port Identifier 128.3.　　//端口的 Cost 值和端口 ID
　　Designated root has priority 4097, address 0023.ac7d.6c80
　　Designated bridge has priority 4097, address 0023.ac7d.6c80

    Designated port id is 128.3, designated path cost 0
    Timers: message age 0, forward delay 0, hold 0
    Number of transitions to forwarding state: 1
    Link type is point-to-point by default
    BPDU: sent 294, received 0   //端口发送和接收的 BPDU 数
  Port 5 (FastEthernet0/3) of VLAN0001 is designated forwarding
    Port path cost 19, Port priority 128, Port Identifier 128.5.
    Designated root has priority 4097, address 0023.ac7d.6c80
    Designated bridge has priority 4097, address 0023.ac7d.6c80
    Designated port id is 128.5, designated path cost 0
    Timers: message age 0, forward delay 0, hold 0
    Number of transitions to forwarding state: 1
    Link type is point-to-point by default
    BPDU: sent 3073, received 0

（4）**show spanning-tree interface f0/22 detail**

S1#**show spanning-tree interface f0/22 detail**
  Port 24 (FastEthernet0/22) of VLAN0001 is designated forwarding
    Port path cost 19, Port priority 128, Port Identifier 128.24.
    Designated root has priority 4097, address 0023.ac7d.6c80
    Designated bridge has priority 4097, address 0023.ac7d.6c80
    Designated port id is 128.24, designated path cost 0
    Timers: message age 0, forward delay 0, hold 0
    Number of transitions to forwarding state: 1
    Link type is point-to-point by default
    BPDU: sent 940, received 1
  Port 24 (FastEthernet0/22) of VLAN0002 is designated forwarding
    Port path cost 19, Port priority 128, Port Identifier 128.24.
    Designated root has priority 4098, address 0023.ac9d.f000
    Designated bridge has priority 28674, address 0023.ac7d.6c80
    Designated port id is 128.24, designated path cost 19
    Timers: message age 0, forward delay 0, hold 0
    Number of transitions to forwarding state: 1
    Link type is point-to-point by default
    BPDU: sent 472, received 1

（5）**show spanning-tree bridge**

S1#**show spanning-tree bridge**

Vlan	Bridge ID	Hello Time	Max Age	Fwd Dly	Protocol
VLAN0001	4097 (4096, 1) 0023.ac7d.6c80	2	20	15	ieee
VLAN0002	28674 (28672, 2) 0023.ac7d.6c80	2	20	15	ieee

//以上显示了生成树网桥信息

3.3 实验 2：Portfast、Uplinkfast 和 Backbonefast

1. 实验目的

通过本实验可以掌握：
① Portfast 的使用场合和配置。
② Uplinkfast 的使用场合和配置。
③ Backbonefast 的使用场合和配置。

2. 实验拓扑

实验拓扑如图 3-11 所示，本实验将会用到路由器 R1。

3. 实验步骤

在实验 1 的基础上继续本实验，我们将只关心 VLAN1 的 STP 树。

（1）配置 Portfast

在图 3-11 中，S1 的 Fa0/1 用于接入路由器 R1。当 R1 接入时，Fa0/1 端口立即进入 Listening 状态，随后经过 Learning，最后才成为 Forwarding，这期间需要 30 秒的时间。这对于有些场合是不可忍受的，可以配置 Portfast 特性，使得设备一接入，端口立即进入 Forwarding 状态。

```
S1(config)#interface fastEthernet0/1
S1(config-if)#switch mode access
S1(config-if)#switch access vlan 1
//默认时，交换机全部端口都在 VLAN1 上，因此上一条命令是可以忽略的
S1(config-if)#spanning-tree portfast
%Warning: portfast should only be enabled on ports connected to a single
  host. Connecting hubs, concentrators, switches, bridges, etc... to this
  interface   when portfast is enabled, can cause temporary bridging loops.
  Use with CAUTION
%Portfast has been configured on FastEthernet0/1but will only
  have effect when the interface is in a non-trunking mode.
```

交换机会警告该端口只能用于接入计算机或者路由器，不要接入其他交换机。在 R1 上打开 G0/0 端口，而在交换机上执行命令"**show spanning-tree vlan 1**"，看端口 Fa0/1 是否立即进入转发状态。

> **提示**
> ① 配置"spanning-tree portfast"属性的端口通常是 Access 口，如果你确信网络中不存在环路，在 Trunk 链路上的端口也可以配置"spanning-tree portfast"属性。全局命令"spanning-tree portfast default"可以把全部端口配置为快速转发，然而请考虑后果。
> ② S1(config-if)#**switchport host**
>   switchport mode will be set to access

spanning-tree portfast will be enabled
channel group will be disabled
//该命令是一条宏命令,也可以把端口配置为 Portfast,同时还配置为 Access 模式,并关闭 EtherChannel 功能

(2)配置 Uplinkfast

先确认实验 1 的 STP 树已经正确收敛。在图 3-11 中的 S1 上,关闭 Fa0/22 口,在 S3 上反复执行"**show spanning-tree vlan 1**"命令,观察 Fa0/2 端口的状态变化。

```
Interface           Role Sts Cost       Prio.Nbr Type
---------------- ---- --- --------- -------- --------------------------------
Fa0/2               Root LIS 19         128.4    P2p
```
大约 15 秒后变为:
```
Fa0/2               Root LRN 19         128.4    P2p
```
大约 15 秒后变为:
```
Fa0/2               Root FWD 19         128.4    P2p
```
合计大约 15+15=30 秒后,Fa0/2 变为转发状态。STP 收敛的时间为 30 秒。

S3(config)#**spanning-tree uplinkfast**   //在交换机 S3 上启用 Uplinkfast 功能

S1(config)#**interface fastEthernet0/22**
S1(config-if)#**no shutdown**
//等 STP 重新稳定后(大约 30 秒),才执行下一语句
S1(config-if)#**shutdown**

在 S3 上重复执行"**show spanning-tree vlan 1**"命令,可以看到 Fa0/2 很快就进入了 Forwarding 状态,STP 重新收敛的时间大大减少。如下所示:

```
S3#show spanning-tree vlan 1
VLAN0001
  Spanning tree enabled protocol ieee
  Root ID    Priority     24577
             Address      0023.ac7d.6c80
             Cost         3019
             Port         3 (FastEthernet0/1)
             Hello Time   2 sec   Max Age 20 sec   Forward Delay 15 sec

  Bridge ID  Priority     49153   (priority 49152 sys-id-ext 1)
```
//配置了 Uplinkfast 后,桥的优先级变为 49152,防止自己成为根桥
```
             Address      0023.ac7d.9c00
             Hello Time   2 sec   Max Age 20 sec   Forward Delay 15 sec
             Aging Time 300
  Uplinkfast enabled

Interface           Role Sts Cost       Prio.Nbr Type
---------------- ---- --- --------- -------- --------------------------------
Fa0/2               Root FWD 3019       128.4    P2p
```
//端口的 Cost 增加了 3000,使得自己不可能成为根桥

技术要点

在没有配置 Uplinkfast 时,如果交换机 S3 能直接检测到 Fa0/1 端口上的链路故障,Fa0/2 会立即进入 Listening 状态,这样 30 秒就能进入 Forwarding 状态。然而如果 S1 和 S3 之间存在一个 Hub,S1 上的 Fa0/22 端口发生故障了,S3 将无法直接检测到故障,S3 只能等待 10 个周期没有收到 S1 的 BPDU(每个周期 2 秒),即 20 秒后,S3 的 Fa0/2 才进入 Listening 状态,这样总共 50 秒才就能进入 Forwarding 状态。所以 STP 重新收敛的时间通常需要 30~50 秒。

```
S3(config)#spanning-tree uplinkfast max-update-rate 200
```

以上命令配置每秒钟所发送的组播包的数目。由于交换机 S3 配置了 Uplinkfast,一旦 Fa0/1 端口有问题,Fa0/2 将快速进入转发状态。但是其他交换机的 MAC 地址表并未更新,交换机 S3 将代理接到根桥设备的 MAC 地址发送组播帧,使得其他交换机更新 MAC 地址表。

```
S3#show spanning-tree uplinkfast
UplinkFast is enabled        //启用了 uplinkfast
Station update rate set to 200 packets/sec    //每秒发送的组播帧数量
UplinkFast statistics    //以下显示 Uplinkfast 统计数
-----------------------
Number of transitions via uplinkFast (all VLANs)          : 1
Number of proxy multicast addresses transmitted (all VLANs) : 4

Name                    Interface List
------------------      ------------------------------------
VLAN0001                Fa0/2(fwd), Fa0/1
```

(3)配置 Backbonefast

打开 S1 上的 Fa0/22 端口,确认 STP 树已经正确收敛。在图 3-11 中的 S1 上,关闭 Fa0/23 和 Fa0/24 端口,在 S3 上反复执行"**show spanning-tree vlan 1**"命令,观察 Fa0/2 端口的状态变化。

```
Interface        Role Sts Cost      Prio.Nbr   Type
---------------- ---- --- ---------- --------- --------
Fa0/2            Desg BLK 3019       128.4     P2p

大约 20 秒后变为:

Fa0/2            Desg LIS 3019       128.4     P2p

大约 15 秒后变为:

Fa0/2            Desg LRN 3019       1284      P2p

大约 15 秒后变为:

Fa0/2            Desg FWD 3019       128.2     P2p

合计大约 20+15+15=50 秒后,Fa0/2 变为转发状态。

S1(config)#spanning-tree backbonefast      //在交换机上启用 Backbonefast 功能
S2(config)#spanning-tree backbonefast      //在交换机上启用 Backbonefast 功能
S3(config)#spanning-tree backbonefast      //在交换机上启用 Backbonefast 功能
S1(config)#interface fastEthernet0/23
S1(config-if)#no shutdown
S1(config-if)#shutdown    //等 STP 重新稳定后,才执行该语句
```

在 S3 上重复执行命令"**show spanning-tree vlan 1**",可以看到 Fa0/2 很快就进入了

Listening 状态，没有了上述 20 秒的等待，用时合计大约 15＋15＝30 秒后，Fa0/2 就变为转发状态，比之前的 50 秒少了 20 秒。

```
S3#show spanning-tree backbonefast
BackboneFast is enabled              //Backbonefast 已经打开
BackboneFast statistics              //以下显示 Backbonefast 统计数
-----------------------
Number of transition via backboneFast (all VLANs)       : 1
Number of inferior BPDUs received (all VLANs)           : 1
Number of RLQ request PDUs received (all VLANs)         : 0
Number of RLQ response PDUs received (all VLANs)        : 1
Number of RLQ request PDUs sent (all VLANs)             : 1
```

技术要点

"**Uplinkfast**"命令只需要在 S3 上配置（接入层交换机）即可，而"**backbonefast**"命令需要在 S1、S2 和 S3 三台交换机上都配置。

3.4 实验 3：RSTP

1. 实验目的

通过本实验可以掌握 RSTP 的配置。

2. 实验拓扑

RSTP 配置实验拓扑如图 3-16 所示。

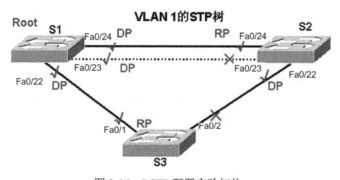

图 3-16 RSTP 配置实验拓扑

3. 实验步骤

在实验 1 和实验 2 的基础上继续本实验，我们将只关心 VLAN1 的 STP 树。

（1）清除实验 2 中的配置

```
S1(config)#no spanning-tree backbonefast
S2(config)#no spanning-tree backbonefast
S3(config)#no spanning-tree backbonefast
```

S3(config)#**no spanning-tree uplinkfast**

确认当前的生成树应该如图 3-16 所示。

（2）配置 RSTP

S1(config)#**spanning-tree mode rapid-pvst**

S2(config)#**spanning-tree mode rapid-pvst**

S3(config)#**spanning-tree mode rapid-pvst**

（3）RSTP 真的快了吗

```
S3#show spanning-tree vlan 1
Interface       Role Sts Cost      Prio.Nbr Type
--------------- ---- --- --------- --------------------------------
Fa0/1           Root FWD 19        128.3    P2p
Fa0/2           Altn BLK 19        128.4    P2p
//以上是 S1 的 f0/22 关闭前的情况
```

在 S1 上关闭 Fa0/22 端口，在 S3 上观察 STP 树的重新生成过程，可以看到 Fa0/2 立即进入了 Forwarding 状态，成为了新的根口。

```
S3#show spanning-tree vlan 1
Interface       Role Sts Cost      Prio.Nbr Type
--------------- ---- --- --------- --------------------------------
Fa0/2           Root FWD 19        128.4    P2p
//以上是 S1 的 Fa0/22 关闭后的情况
```

在 S1 上重新打开 Fa0/22 端口，确认 STP 稳定后，在 S2 上关闭 Fa0/23 和 Fa0/24 两个端口，在 S3 上重复执行"**show spanning-tree vlan 1**"命令，可以看到 Fa0/2 立即进入 Forwarding 状态，说明 RSTP 的收敛速度比普通 STP 有了很大提高。

（4）配置链路类型

S1(config)#**interface range fastEthernet0/23 -24**

S1(config-if-range)#**duplex full**

S1(config-if-range)#**spanning-tree link-type point-to-point**

S2(config)#**interface range fastEthernet0/23 -24**

S2(config-if-range)#**duplex full**

S2(config-if-range)#**spanning-tree link-type point-to-point**

//S1 和 S2 之间的链路是 Trunk 链路，自动协商为全双工，RSTP 会自动把它们的链路类型标识为点到点。这里强制配置了一遍

```
S1#show spanning-tree vlan 1 interface fastEthernet0/23 detail
Port 25 (FastEthernet0/23) of VLAN0001 is designated forwarding
   Port path cost 19, Port priority 128, Port Identifier 128.25.
   Designated root has priority 4097, address 0023.ac7d.6c80
   Designated bridge has priority 4097, address 0023.ac7d.6c80
   Designated port id is 128.25, designated path cost 0
   Timers: message age 0, forward delay 0, hold 0
   Number of transitions to forwarding state: 1
```

Link type is point-to-point　　//链路的类型为点对点
　　BPDU: sent 21, received 1

📖 技术要点

在 RSTP 中端口分为边界端口（Edge Port）、点到点端口（Point-to-Point Port）和共享端口（Share Port）。如果端口上配置了 spanning-tree portfast，端口就为边界端口；如果端口是半双工模式，端口就为共享端口；如果端口是全双工模式，端口就为点到点端口。在端口上指明类型有利于 RSTP 的运行。

4. 实验调试

（1）show spanning-tree summary

```
S1#show spanning-tree summary
Switch is in rapid-pvst mode    //启用 RSTP
Root bridge for: VLAN0001
Extended system ID            is enabled
Portfast Default              is disabled
PortFast BPDU Guard Default   is disabled
Portfast BPDU Filter Default  is disabled
Loopguard Default             is disabled
EtherChannel misconfig guard is enabled
UplinkFast                    is disabled
BackboneFast                  is disabled
Configured Pathcost method used is short
```

Name	Blocking	Listening	Learning	Forwarding	STP Active
VLAN0001	0	0	0	7	7
VLAN0002	1	0	0	2	3
2 vlans	1	0	0	9	10

以上显示了各 VLAN 有多少端口处于不同的状态。

（2）show spanning-tree root

```
S1#show spanning-tree root
```

Vlan	Root ID	Cost	Time	Age	Dly	Root Port
VLAN0001	4097 0023.ac7d.6c80	0	2	20	15	
VLAN0002	4098 0023.ac9d.f000	19	2	20	15	Fa0/23

以上显示了各个 VLAN 的根桥情况。Root ID 列：根桥的 ID；Cost：到根桥的 Cost 值；Time 列：Hello 时间；Age：最大老化时间；Dly：转发延时；Root Port：根口。

3.5 实验 4：MSTP

1. 实验目的

通过本实验可以掌握：
① MSTP 的工作原理。
② MSTP 的配置。
③ MSTP 和 PVST 的兼容性。

2. 实验拓扑

MSTP 实验拓扑如图 3-17 所示。交换机 S1 和 S2 将使用 MSTP 协议，而交换机 S3 使用 PVST。

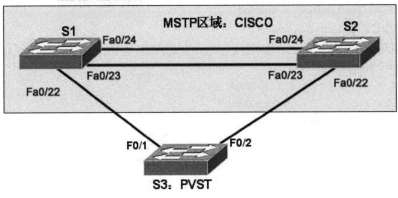

图 3-17　MSTP 实验拓扑

3. 实验步骤

（1）准备工作

Trunk 的配置：如图 3-17 所示，把 S1 和 S2 的 Fa0/22～Fa0/24 端口配置为 Trunk 端口，S3 的 Fa0/1～Fa0/2 端口配置为 Trunk 端口。

VLAN 的配置：如图 3-17 所示，在 S1、S2 和 S3 上创建 VLAN2、VLAN3 和 VLAN4。VLAN1 和 VLAN2 将使用 MST 实例 1，VLAN3 和 VLAN4 将使用 MST 实例 2。

（2）配置 MSTP

```
S1(config)#spanning-tree mode mst
//把生成树的模式改为 MSTP，默认时是 PVST+
S1(config)#spanning-tree mst configuration    //进入 MSTP 的配置模式
S1(config-mst)#name CISCO    //命名 MSTP 的名字
```

```
S1(config-mst)#revision 1    //配置 MST 的 Revision 号,只有名字和 Revision 号相同的交换机才是在同
一个 MST 区域的交换机
S1(config-mst)#instance 1 vlan 1-2    //把 VLAN1 和 VLAN2 的生成树映射到实例 1
S1(config-mst)#instance 2 vlan 3-4    //把 VLAN3 和 VLAN4 的生成树映射到实例 2,这里一共有 3 个
MST 实例。实例 0 是默认的实例,默认时所有的 VLAN 都映射到该实例
S1(config-mst)#exit    //要退出,配置才能生效
S1(config)#spanning-tree mst 1 priority 8192    //配置 S1 为 MST 实例 1 的根桥
S1(config)#spanning-tree mst 2 priority 12288


S2(config)#spanning-tree mode mst
S2(config)#spanning-tree mst configuration
S2(config-mst)#name CISCO
S2(config-mst)#revision 1
S2(config-mst)#instance 1 vlan 1-2
S2(config-mst)#instance 2 vlan 3-4
S2(config-mst)#exit
S2(config)#spanning-tree mst 1 priority 12288
S2(config)#spanning-tree mst 2 priority 8192    //配置 S2 为 MST 实例 2 的根桥


S3(config)#spanning-tree mode pvst    //配置 S3 运行 PVST,而不是 MSTP
S3(config)#spanning-tree vlan 1-4 priority 24576
```

4. 实验调试

(1) show spanning-tree

```
S1#show spanning-tree mst
##### MST0    vlans mapped:    5-4094
//实例 0 的生成树,默认时所有的 VLAN 都在实例 0 上,该实例就是 CIST
Bridge        address 0023.ac7d.6c80   priority      32768 (32768 sysid 0)
Root          address 0023.ac7d.9c00   priority      24577 (24576 sysid 1)
//以上是总根(S3 交换机)的信息,
              port    Fa0/22         path cost      200000
Regional Root this switch    //该交换机是 MST0 的区域根
Operational   hello time 2 , forward delay 15, max age 20, txholdcount 6
Configured    hello time 2 , forward delay 15, max age 20, max hops   20
Interface           Role Sts Cost         Prio.Nbr Type
--------------- ---- --- --------- -------- --------------------------------
Fa0/1               Desg FWD 200000    128.3     P2p Edge
Fa0/3               Desg FWD 200000    128.5     P2p
Fa0/4               Desg FWD 200000    128.6     P2p
Fa0/10              Desg FWD 200000    128.12    P2p
Fa0/22              Root BKN*200000    128.24    P2p Bound(PVST) *PVST_Inc
//该端口是边缘端口,端口目前处于 Broken 状态,有问题稍后解决。此外 MST 中 100 Mbps 端口的 Cost
```

为 200000

```
    Fa0/23          Desg FWD 200000    128.25      P2p
    Fa0/24          Desg FWD 200000    128.26      P2p

##### MST1       vlans mapped:    1-2    //MST1 实例的生成树信息
Bridge           address 0023.ac7d.6c80  priority      8193   (8192 sysid 1)
Root             this switch for MST1    //该交换机是 MST1 的区域根
Interface        Role Sts Cost       Prio.Nbr Type
---------------- ---- --- --------- -------- --------------------------------
Fa0/1            Desg FWD 200000     128.3    P2p Edge
Fa0/3            Desg FWD 200000     128.5    P2p
Fa0/4            Desg FWD 200000     128.6    P2p
Fa0/10           Desg FWD 200000     128.12   P2p
Fa0/22           Mstr BKN*200000     128.24   P2p Bound(PVST) *PVST_Inc
Fa0/23           Desg FWD 200000     128.25   P2p
Fa0/24           Desg FWD 200000     128.26   P2p

##### MST2       vlans mapped:    3-4    //MST2 实例的生成树信息
Bridge           address 0023.ac7d.6c80  priority      12290 (12288 sysid 2)
Root             address 0023.ac9d.f000  priority      8194  (8192 sysid 2)
                 port    Fa0/23          cost          200000        rem hops 19

Interface        Role Sts Cost       Prio.Nbr Type
---------------- ---- --- --------- -------- --------------------------------
Fa0/22           Mstr BKN*200000     128.24   P2p Bound(PVST) *PVST_Inc
Fa0/23           Root FWD 200000     128.25   P2p
Fa0/24           Altn BLK 200000     128.26   P2p
```

（2）**show spanning-tree mst configuration**

该命令用来查看 MST 的配置。

```
S1#show spanning-tree mst configuration
Name       [CISCO]
Revision   1      Instances configured 3
Instance   Vlans mapped
--------   -------------------------------------------------------------------
0          5-4094
1          1-2
2          3-4
-------------------------------------------------------------------------------
//以上显示各 MST 实例对应的 VLAN
```

📖 **技术要点**

MST 完成配置后，对 MST 树的控制方法和 PVST 是类似的，我们同样可以通过控制不同实例的 MST 树来达到负载平衡等目的，这里不再赘述。

第 3 章　STP

（3）解决 S1 上的 Fa0/22 状态为 Broken 问题

当 MST 和 PVST 结合时，如果 PVST 中的交换机声称自己为 VLAN 的根，则 MST 的 CIST 会阻断到根的上连链路。解决办法是：把 MST 中的交换机配置为 VLAN 的根。如下所示：

S1(config)#**spanning-tree mst 0 root primary**

在 MST 中，MST0 就是 CIST，所以配置 MST0 的优先级即可。

S1(config)#
*Mar 1 13:37:41.183: %SPANTREE-2-PVSTSIM_OK: PVST Simulation inconsistency cleared on port FastEthernet0/22.
//S1 的 Fa0/22 端口恢复正常了

```
S1#show spanning-tree mst
##### MST0      vlans mapped:    5-4094
Bridge          address 0023.ac7d.6c80  priority      24576 (24576 sysid 0)
Root            this switch for the CIST    //该根桥为总根
Operational     hello time 2 , forward delay 15, max age 20, txholdcount 6
Configured      hello time 2 , forward delay 15, max age 20, max hops     20

Interface       Role Sts Cost      Prio.Nbr Type
---------------- ---- --- --------- --------------------------------
Fa0/1           Desg FWD 200000    128.3    P2p Edge
Fa0/3           Desg FWD 200000    128.5    P2p
Fa0/4           Desg FWD 200000    128.6    P2p
Fa0/10          Desg FWD 200000    128.12   P2p
Fa0/22          Desg FWD 200000    128.24   P2p Bound(PVST)   //边界端口，现在正常了
Fa0/23          Desg FWD 200000    128.25   P2p
Fa0/24          Desg FWD 200000    128.26   P2p

##### MST1      vlans mapped:    1-2
Bridge          address 0023.ac7d.6c80  priority      8193  (8192 sysid 1)
Root            this switch for MST1

Interface       Role Sts Cost      Prio.Nbr Type
---------------- ---- --- --------- --------------------------------
Fa0/1           Desg FWD 200000    128.3    P2p Edge
Fa0/3           Desg FWD 200000    128.5    P2p
Fa0/4           Desg FWD 200000    128.6    P2p
Fa0/10          Desg FWD 200000    128.12   P2p
Fa0/22          Desg FWD 200000    128.24   P2p Bound(PVST)
Fa0/23          Desg FWD 200000    128.25   P2p
Fa0/24          Desg FWD 200000    128.26   P2p
         ……………… (略)
```

（4）**show spanning-tree mst interface f0/22 detail**

S1#**show spanning-tree mst interface f0/22 detail**

```
Edge port: no            (default)         port guard : none      (default)
Link type: point-to-point (auto)           bpdu filter: disable   (default)
Boundary : boundary      (PVST)            bpdu guard : disable   (default)

FastEthernet0/22 of MST0 is designated forwarding     //MST0 中该端口的状态
Vlans mapped to MST0 5-4094
Port info           port id         128.24   priority    128    cost     200000
Designated root     address 0023.ac7d.6c80   priority   24576   cost          0
Design. regional root address 0023.ac7d.6c80 priority   24576   cost          0
Designated bridge   address 0023.ac7d.6c80   priority   24576   port id  128.24
Timers: message expires in 0 sec, forward delay 0, forward transitions 7
Bpdus sent 61203, received 15760

FastEthernet0/22 of MST1 is designated forwarding
Vlans mapped to MST1 1-2
Port info           port id         128.24   priority    128    cost     200000
Designated root     address 0023.ac7d.6c80   priority    8193   cost          0
Designated bridge   address 0023.ac7d.6c80   priority    8193   port id  128.24
Timers: message expires in 0 sec, forward delay 0, forward transitions 9
Bpdus (MRecords) sent 61203, received 0

FastEthernet0/22 of MST2 is designated forwarding
Vlans mapped to MST2 3-4
Port info           port id         128.24   priority    128    cost     200000
Designated root     address 0023.ac9d.f000   priority    8194   cost     200000
Designated bridge   address 0023.ac7d.6c80   priority   12290   port id  128.24
Timers: message expires in 0 sec, forward delay 0, forward transitions 9
Bpdus (MRecords) sent 61207, received 0
```

3.6 实验 5：STP 树保护

1. 实验目的

通过本实验可以掌握：
① Guard Root 的使用方法。
② BPDU Guard 的使用方法。
③ BPDU Filter 的使用方法。

2. 实验拓扑

STP 树保护实验拓扑如图 3-18 所示。

第 3 章 STP

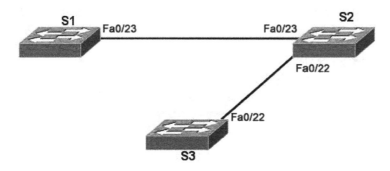

图 3-18 STP 树保护实验拓扑

3. 实验步骤

（1）关闭不需要的端口，配置 S1、S2 和 S3 之间的 Trunk

```
S1(config)#interface fastEthernet0/22
S1(config-if)#shutdown
S1(config)#interface fastEthernet0/24
S1(config-if)#shutdown
S1(config)#interface fastEthernet0/23
S1(config-if)#switchport trunk encapsulation dot1q
S1(config-if)#switchport mode trunk

S2(config)#interface fastEthernet0/22
S2(config-if)#switchport trunk encapsulation dot1q
S2(config-if)#switchport mode trunk
S2(config-if)#interface fastEthernet0/23
S2(config-if)#switchport trunk encapsulation dot1q
S2(config-if)#switchport mode trunk
S2(config-if)#interface fastEthernet0/24
S2(config-if)#shutdown

S3(config)#interface fastEthernet0/1
S3(config-if)#shutdown
S3(config)#interface fastEthernet0/2
S3(config-if)#switchport trunk encapsulation dot1q
S3(config-if)#switchport mode trunk
```

（2）配置 S1 成为根桥

```
S1(config)#spanning-tree vlan 1 priority 8192
```

（3）在 S2 的 Fa0/22 上配置 Guard Root

```
S2(config)#interface fastEthernet0/22
S2(config-if)#spanning-tree guard root
```

(4) 把 S3 改为根桥（S3 模拟非法交换机接入到网络中，它想成为新的根桥），观察 S2 的动作

```
S3(config)#spanning-tree vlan 1 priority 4096
```
在 S2 交换机上应会有错误信息弹出，如下所示：

*Mar 1 00:47:07.824: %SPANTREE-2-ROOTGUARD_CONFIG_CHANGE: Root guard enabled on port FastEthernet0/22.

*Mar 1 00:47:23.335: %SPANTREE-2-ROOTGUARD_BLOCK: Root guard blocking port FastEthernet0/22 on VLAN0001.

```
S2#show spanning-tree inconsistentports
Name                 Interface                Inconsistency
-------------------- ------------------------ ------------------
VLAN0001             FastEthernet0/22         Root Inconsistent
Number of inconsistent ports (segments) in the system : 1
```
//S2 将从 Fa0/22 收到 S3 发送的更优的 BPDU，然而由于该端口上配置了 Guard Root，S2 的端口进入阻断状态

```
S2#show spanning-tree
VLAN0001
（此处省略）
Interface        Role Sts Cost      Prio.Nbr Type
---------------- ---- --- --------- -------- --------------------------------
Fa0/22           Desg BKN*19        128.24   P2p *ROOT_Inc
Fa0/23           Root FWD 19        128.25   P2p

S2#show spanning-tree vlan 1 interface f0/22 detail
 Port 24 (FastEthernet0/22) of VLAN0001 is broken    (Root Inconsistent)
   Port path cost 19, Port priority 128, Port Identifier 128.24.
   Designated root has priority 8193, address 0023.ac7d.6c80
   Designated bridge has priority 32769, address 0023.ac9d.f000
   Designated port id is 128.24, designated path cost 19
   Timers: message age 2, forward delay 0, hold 0
   Number of transitions to forwarding state: 1
   Link type is point-to-point by default
   Root guard is enabled on the port
   BPDU: sent 1142, received 389
```

技术要点

配 Guard Root 功能是为了防止用户擅自在网络中接入交换机并成为新的根桥，从而破坏了原有的 STP 树。该功能通常在接入层交换机上对外开放的端口上配置，这些端口将拒绝接收比现有根桥更优的 BPDU。当威胁消除后，端口将自动恢复正常。

(5) 配置 BPDU Guard

```
S3(config)#interface fastEthernet0/2
```

第 3 章　STP

```
S3(config-if)#switchport mode access    //端口改为 Access 模式

S2(config)#interface fastEthernet0/22
S2(config-if)#shutdown    //关闭端口
S2(config-if)#no spanning-tree guard root    //去掉之前的配置
S2(config-if)#switchport mode access    //端口改为 Access 模式
S2(config-if)#spanning-tree portfast
S2(config-if)#spanning-tree bpduguard enable    //配置 BPDU Guard
S2(config)#interface fastEthernet0/22
S2(config-if)#no shutdown
*Mar  1 01:01:08.237: %SPANTREE-2-BLOCK_BPDUGUARD: Received BPDU on port Fa0/22 with BPDU Guard enabled. Disabling port.
*Mar  1 01:01:08.237: %PM-4-ERR_DISABLE: bpduguard error detected on Fa0/22, putting Fa0/22 in err-disable state
*Mar  1 01:01:09.244: %LINEPROTO-5-UPDOWN: Line protocol on Interface FastEthernet0/22, changed state to down
```
//交换机 S2 从 Fa0/22 端口收到 S3 的 BPDU，Fa0/22 被关闭了。S2 的 Fa0/22 应该接入不会发送 BPDU 的计算机，然而现在接入了一台会发送 BPDU 的交换机 S3

```
S2#show interfaces FastEthernet0/22
FastEthernet0/22 is down, line protocol is down (err-disabled)
```
//可以看到，Fa0/22 端口被关闭了，要重新开启，请先移除 BPDU 源，在该端口执行 "**shutdown**" 和 "**no shutdown**" 命令

```
S2#show spanning-tree interface f0/22 detail
 Port 24 (FastEthernet0/22) of VLAN0001 is designated forwarding
   Port path cost 19, Port priority 128, Port Identifier 128.24.
   Designated root has priority 8193, address 0023.ac7d.6c80
   Designated bridge has priority 32769, address 0023.ac9d.f000
   Designated port id is 128.24, designated path cost 19
   Timers: message age 0, forward delay 0, hold 0
   Number of transitions to forwarding state: 1
   The port is in the portfast mode
   Link type is point-to-point by default
   Bpdu guard is enabled
   BPDU: sent 76, received 0
```

技术要点

① 可以配置端口自动恢复，如下所示：

```
S2(config)#errdisable recovery cause bpduguard
```
//允许因为 bpduguard 而关闭的端口排除故障后自动恢复
```
S2(config)#errdisable recovery interval 60
```
//配置自动恢复的时间为 60 秒

② 配置 BPDU Guard 功能是为了防止在那些已经配置 portfast 命令的端口上接入交换

机,从而导致环路的产生。因为 portfast 端口一激活就立即进入转发状态,这些端口通常用于接入计算机。BPDU Guard 功能可以防止这些端口收到 BPDU。

(6) 配置 BPDU Filter

```
S2(config)#default interface f0/22
//把端口的配置恢复为默认配置
S2(config)#interface FastEthernet0/22
S2(config-if)#switchport mode access
S2(config-if) #spanning-tree portfast
//配置端口为 portfast,先不启用 bpdufilter

S3(config)#spanning-tree vlan 1 priority 16384
S3#debug spanning-tree bpdu
//把 S3 的优先级降低,并把 BPDU 的 debug 打开。这时根桥是 S1,在 S3 上应该能够看到它在接收 BPDU

S2(config)#default interface f0/22
S2(config-if)#spanning-tree bpdufilter enable
//在 Fa0/22 端口上启用 bpdufilter,这时 S3 将不再接收到 S2 转发的 BPDU 了。经过 20 秒后,S3 将认
为自己是根桥了。如下所示
S3#show spanning-tree vlan 1
VLAN0001
  Spanning tree enabled protocol ieee
  Root ID    Priority    16385
             Address     0023.ac7d.9c00
             This bridge is the root   //由于收不到 BPDU,S3 认为自己就是根桥了
             Hello Time   2 sec  Max Age 20 sec  Forward Delay 15 sec
  Bridge ID  Priority    16385  (priority 16384 sys-id-ext 1)
             Address     0023.ac7d.9c00
             Hello Time   2 sec  Max Age 20 sec  Forward Delay 15 sec
             Aging Time 15
Interface         Role Sts Cost      Prio.Nbr   Type
----------------- ---- --- --------- --------- -----------------------
Fa0/2             Desg FWD 19        128.4     P2P
```

 提示

① 也可以在全部配置模式下配置 bpdufilter,使用的命令为"**spanning-tree portfast bpdufilter default**"。在全局下配置 bpdufilter,如果端口仅配置了 portfast,而没有配置 bpdufilter,一旦该端口收到 BPDU,则该端口将丢弃 portfast 和 bpdufilter 状态,正常收发 BPDU。

② 如果在端口上同时配置 bpdufilter 和 bpduguard,因为 bpdufilter 的优先级高,bpduguard 将不起作用。

3.7 实验 6：环路防护

1. 实验目的

通过本实验可以掌握：
① Guard Loop 的使用方法。
② UDLD 的使用方法。

2. 实验拓扑

环路保护实验拓扑如图 3-19 所示。注意：我们在 S2 和 S3 之间加入了一个 Hub（不能是交换机）。

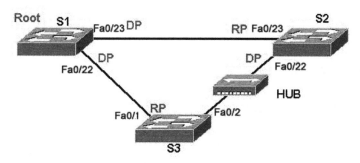

图 3-19 环路保护实验拓扑

3. 实验步骤

（1）准备工作

配置 S1、S2 和 S3 之间的 Trunk，如图 3-19 所示配置 STP 树。

```
S1(config)#interface fastEthernet0/22
S1(config-if)#switchport trunk encapsulation dot1q
S1(config-if)#switchport mode trunk
S1(config)#interface fastEthernet0/23
S1(config-if)#switchport trunk encapsulation dot1q
S1(config-if)#switchport mode trunk
S1(config)#interface fastEthernet0/24
S1(config-if)#shutdown
S1(config)# spanning-tree vlan 1 priority 4096

S2(config)#interface fastEthernet0/22
S2(config-if)#switchport trunk encapsulation dot1q
S2(config-if)#switchport mode trunk
S2(config)#interface fastEthernet0/23
S2(config-if)#switchport trunk encapsulation dot1q
S2(config-if)#switchport mode trunk
```

```
S2(config)#interface fastEthernet0/24
S2(config-if)#shutdown
S2(config)# spanning-tree vlan 1 priority 8192

S3(config)#interface fastEthernet0/1
S3(config-if)#switchport trunk encapsulation dot1q
S3(config-if)#switchport mode trunk
S3(config)#interface fastEthernet0/2
S3(config-if)#switchport trunk encapsulation dot1q
S3(config-if)#switchport mode trunk
```

（2）配置 Guard Loop

```
S3(config)#interface fastEthernet0/2
S3(config-if)#spanning-tree guard loop


S3#show spanning-tree vlan 1
VLAN0001
  Spanning tree enabled protocol ieee
  Root ID    Priority    4097
             Address     0023.ac7d.6c80
             Cost        19
             Port        3 (FastEthernet0/1)
             Hello Time  2 sec   Max Age 20 sec   Forward Delay 15 sec
  Bridge ID  Priority    32769   (priority 32768 sys-id-ext 1)
             Address     0023.ac7d.9c00
             Hello Time  2 sec   Max Age 20 sec   Forward Delay 15 sec
             Aging Time 15

Interface           Role Sts Cost      Prio.Nbr Type
------------------- ---- --- --------- -------- --------------------------------
Fa0/1               Root FWD 19        128.3    P2p
Fa0/2               Altn BLK 100       128.4    Shr    //该端口是替代端口
```

在 S2 上关闭 Fa0/22 端口，模拟单向故障（S2 无法发送数据包到 S3），则 20 秒后（最大寿命计时器），在 S3 上出现：

```
*Mar  1 04:35:04.728: %SPANTREE-2-LOOPGUARD_BLOCK: Loop guard blocking port FastEthernet0/2 on VLAN0001.


S3#show spanning-tree vlan 1
VLAN0001
  Spanning tree enabled protocol ieee
  Root ID    Priority    4097
             Address     0023.ac7d.6c80
             Cost        19
             Port        3 (FastEthernet0/1)
             Hello Time  2 sec   Max Age 20 sec   Forward Delay 15 sec
```

第 3 章　STP

```
    Bridge ID   Priority      32769    (priority 32768 sys-id-ext 1)
                Address       0023.ac7d.9c00
                Hello Time    2 sec   Max Age 20 sec   Forward Delay 15 sec
                Aging Time 300
Interface              Role Sts Cost      Prio.Nbr Type
------------------- ---- --- --------- -------- --------------------------------
Fa0/1                  Root FWD 19         128.3    P2p
Fa0/2                  Desg BKN*100        128.4    Shr *LOOP_Inc
```
//由于该端口配置了 guard loop，50 秒（20+15+15）后处于 broken 状态，而不是进入转发状态

```
S3#show spanning-tree interface f0/2 detail
 Port 4 (FastEthernet0/2) of VLAN0001 is broken    (Loop Inconsistent)
   Port path cost 100, Port priority 128, Port Identifier 128.4.
   Designated root has priority 4097, address 0023.ac7d.6c80
   Designated bridge has priority 32769, address 0023.ac7d.9c00
   Designated port id is 128.4, designated path cost 19
   Timers: message age 0, forward delay 0, hold 0
   Number of transitions to forwarding state: 1
   Link type is shared by default
   Loop guard is enabled on the port
   BPDU: sent 87, received 96
```

（3）配置 UDLD

先把步骤（2）的配置清除：

```
S3(config)#int fastEthernet 0/2
S3(config-if)#no spanning-tree guard loop
S2(config-if)#int fastEthernet 0/2
S2(config-if)#no shutdown
```

配置 UDLD：

```
S2(config)#udld aggressive
```
//配置 UDLD 模式为主动模式，这样检测到单向链路故障后，将关闭端口。"udld enable"命令则是配置 UDLD 为普通模式
```
  S2(config)#int fastEthernet 0/22
  S2(config-if)#udld port aggressive


  S3(config)#udld aggressive
  S3(config)#udld message time 15
```
//配置每 15 秒发送一次探测包，默认值就是 15 秒
```
  S3(config)#int fastEthernet 0/2
  S3(config-if)#udld port aggressive


S3#show udld neighbors
Port       Device Name     Device ID        Port ID      Neighbor State
----       -----------     ---------        -------      --------------
```

Fa0/2	FDO1243Y2FG	1		Fa0/22	Bidirectional

//在 S3 可以看到 UDLD 邻居。只有链路的两端都配置了 UDLD 才能建立邻居关系

技术要点

只有链路的两端都配置了 UDLD 才能建立邻居关系，UDLD 发现原有的邻居不在了才认为发生了单向故障。如果一端配置了 UDLD，另一端并没有配置 UDLD，是无法检测单向链路故障的。

```
S3#show udld f0/2
Interface Fa0/2
---
Port enable administrative configuration setting: Enabled / in aggressive mode
Port enable operational state: Enabled / in aggressive mode
Current bidirectional state: Bidirectional    //当前链路的状态是双向的，正常
Current operational state: Advertisement - Single neighbor detected
Message interval: 15    //每 15 秒发送一次探测包
Time out interval: 5


    Entry 1
    ---
    Expiration time: 43
    Device ID: 1
    Current neighbor state: Bidirectional
    Device name: FDO1243Y2FG
    Port ID: Fa0/22
    Neighbor echo 1 device: FDO1243Y2BJ
    Neighbor echo 1 port: Fa0/2

    Message interval: 15
    Time out interval: 5
    CDP Device name: S2
        // UDLD 邻居的信息

S3#show spanning-tree vlan 1
VLAN0001
  Spanning tree enabled protocol ieee
  Root ID    Priority    4097
             Address     0023.ac7d.6c80
             Cost        19
             Port        3 (FastEthernet0/1)
             Hello Time  2 sec  Max Age 20 sec   Forward Delay 15 sec

  Bridge ID  Priority    32769  (priority 32768 sys-id-ext 1)
             Address     0023.ac7d.9c00
             Hello Time   2 sec   Max Age 20 sec   Forward Delay 15 sec
```

```
                     Aging Time 300
Interface            Role Sts Cost      Prio.Nbr Type
------------------- ---- --- --------- --------------------------------
Fa0/1                Root FWD 19        128.3    P2p
Fa0/2                Altn BLK 100       128.4    Shr
```
//命令检查 VLAN1 的 STP 情况，S3 的 Fa0/2 处于 blocking 状态，该端口从 S2 收到 BPDU

//命令关闭 S2 的 Fa0/22 端口，模拟 S2 和 S3 间的单向故障（S3 将不能从 S2 收到 BPDU 了）
S2(config)#**int fastEthernet 0/22**
S2(config-if)#**shutdown**

//几十秒后，在 S3 交换机上将出现如下信息：
*Mar 1 04:51:48.391: %UDLD-4-UDLD_PORT_DISABLED: UDLD disabled interface Fa0/2, aggressive mode failure detected
*Mar 1 04:51:48.391: %PM-4-ERR_DISABLE: udld error detected on Fa0/2, putting Fa0/2 in err-disable state   //检测到单向故障
*Mar 1 04:51:49.398: %LINEPROTO-5-UPDOWN: Line protocol on Interface FastEthernet0/2, changed state to down   //关闭 Fa0/22 端口
*Mar 1 04:51:50.405: %LINK-3-UPDOWN: Interface FastEthernet0/2, changed state to down

S3#**show interfaces f0/2**
FastEthernet0/2 is down, line protocol is down (err-disabled)   //端口被关闭
  Hardware is Fast Ethernet, address is 0023.ac7d.9c04 (bia 0023.ac7d.9c04)
  MTU 1500 bytes, BW 10000 Kbit, DLY 1000 usec,
     reliability 255/255, txload 1/255, rxload 1/255
  Encapsulation ARPA, loopback not set

技术要点

排除单向故障后，可以使用 "**udld reset**" 命令来使端口重新可用。当然也可以使用 "**errdisable recovery cause udld**" 全局命令让端口自动重新可用。

3.8 实验 7：FlexLink

1. 实验目的

通过本实验可以掌握 FlexLink 的使用方法。

2. 实验拓扑

FlexLink 实验拓扑如图 3-20 所示。S3 充当接入层交换机，S1 和 S2 充当分布层或核心层交换机。

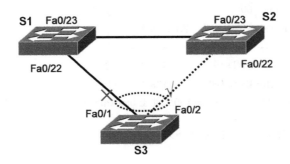

图 3-20　FlexLink 实验拓扑

3. 实验步骤

（1）准备工作

配置 S1、S2 和 S3 之间的 Trunk，如图 3-20 所示配置 STP 树。

```
S1(config)#interface fastEthernet0/22
S1(config-if)#switchport trunk encapsulation dot1q
S1(config-if)#switchport mode trunk
S1(config)#interface fastEthernet0/23
S1(config-if)#switchport trunk encapsulation dot1q
S1(config-if)#switchport mode trunk
S1(config)#interface fastEthernet0/24
S1(config-if)#shutdown
S1(config)# spanning-tree vlan 1 priority 4096


S2(config)#interface fastEthernet0/22
S2(config-if)#switchport trunk encapsulation dot1q
S2(config-if)#switchport mode trunk
S2(config)#interface fastEthernet0/23
S2(config-if)#switchport trunk encapsulation dot1q
S2(config-if)#switchport mode trunk
S2(config)#interface fastEthernet0/24
S2(config-if)#shutdown
S2(config)# spanning-tree vlan 1 priority 8192


S3(config)#interface fastEthernet0/1
S3(config-if)#switchport mode access
S3(config-if)#switchport access vlan 1
S3(config)#interface fastEthernet0/2
S3(config-if)#switchport mode access
S3(config-if)#switchport access vlan 1


S1(config)#interface Vlan1
```

第 3 章 STP

S1(config-if)#**ip address 192.168.1.10 255.255.255.0**

S3(config)#**interface Vlan1**
S3(config-if)#**ip address 192.168.1.30 255.255.255.0**

（2）配置 FlexLink

S3(config)#**interface FastEthernet0/1**
S3(config-if)#**switchport backup interface Fa0/2**
//配置 Fa0/2 是 Fa0/1 端口的备份端口
S3(config-if)#**switchport backup interface Fa0/2 preemption mode forced**
//配置抢占模式。preemption mode 有 forced、bandwith 和 off 三种，默认是 off。当主用端口发生故障而备用端口激活后，在主用端口恢复正常后，如果 preemption mode 是 off，则主用端口不会激活；如果 preemption mode 是 forced，则主用端口会激活；如果 preemption mode 是 bandwidth，则带宽大的端口会激活
S3(config-if)#**switchport backup interface Fa0/2 preemption delay 30**
//配置抢占延时，单位为秒

S3(config)#**mac address-table move update transmit**
S3(config)#**interface FastEthernet0/1**
S3(config-if)#**switchport backup interface Fa0/2 mmu primary vlan 1**
//配置"MAC Address-Table Move Update"，当主用端口发生故障后，备用端口启用，这时交换机 S1 和 S2 的 MAC 地址表并没有及时更新。配置"MAC Address-Table Move Update"后，交换机 S3 将发送更新消息给 S1 和 S2
S1(config)#**mac address-table move update receive**
S2(config)#**mac address-table move update receive**
//"MAC Address-Table Move Update"功能不仅需要在接入层交换机上配置，也需要在分布层交换机上配置，S3 发送信息，S1 和 S2 接收信息

4. 实验调试

（1）测试 FlexLink

从 S3 上连续 ping S1(192.168.1.10)，然后在 S1 上关闭 Fa0/22 端口，如下所示：
S3#**ping 192.168.1.30 repeat 10000000**
S1(config)# **interface FastEthernet 0/22**
S1(config-if)#**shutdown**
在 S3 应该看到通信是连续的。

（2）show interfaces switchport backup detail

```
S3#show interfaces switchport backup detail
Switch Backup Interface Pairs:
Active Interface        Backup Interface        State
------------------------------------------------------------
FastEthernet0/1         FastEthernet0/2         Active Down/Backup Up
```
//可以看到 Fa0/1 是主用端口（Active），但当前状态是 Down，Fa0/2 是备用端口（Backup），但当前状态是 Up

Preemption Mode: forced   //抢占模式
Preemption Delay : 30 seconds   //抢占延时
Multicast Fast Convergence: Off
Bandwidth : 100000 Kbit (Fa0/1), 100000 Kbit (Fa0/2)   //端口的带宽
Mac Address Move Update Vlan : 1   // Mac Address Move Update 发送哪个 VLAN

（3）show mac address-table move update

S3#**show mac address-table move update**
Switch-ID : 0123.ac7d.9c00
Dst mac-address : 0180.c200.0010
Vlans/Macs supported : 1023/6272
Default/Current settings: Rcv Off/Off, Xmt Off/On
Max packets per min : Rcv 40, Xmt 60

Rcv packet count : 0
Rcv conforming packet count : 0
Rcv invalid packet count : 0
Rcv packet count this min : 0
Rcv threshold exceed count : 0
Rcv last sequence# this min : 0
Rcv last interface : None
Rcv last src-mac-address : 0000.0000.0000
Rcv last switch-ID : 0000.0000.0000

Xmt packet count : 4   // Mac Address Move Update 发送的数量
Xmt packet count this min : 0
Xmt threshold exceed count : 0
Xmt pak buf unavail cnt : 0
Xmt last interface : Fa0/2

S1#**show mac address-table move update**
Switch-ID : 0123.ac7d.6c80
Dst mac-address : 0180.c200.0010
Vlans/Macs supported : 1023/6272
Default/Current settings: Rcv Off/On, Xmt Off/Off
Max packets per min : Rcv 40, Xmt 60

Rcv packet count : 4   // Mac Address Move Update 接收的数量
Rcv conforming packet count : 4
Rcv invalid packet count : 0
Rcv packet count this min : 0
Rcv threshold exceed count : 0
Rcv last sequence# this min : 0
Rcv last interface : Fa0/23

Rcv last src-mac-address : 0023.ac7d.9c04
Rcv last switch-ID : 0123.ac7d.9c00

Xmt packet count : 0
Xmt packet count this min : 0
Xmt threshold exceed count : 0
Xmt pak buf unavail cnt : 0
Xmt last interface : None

3.9 本章小结

本章首先介绍了 STP 的作用和基本工作原理，交换机通过 STP 协议有选择性地阻断了某些端口，从而构建无环路的转发路径，STP 需要选举根桥、根口和指定口。IEEE 802.1d 的 STP 需要较长时间才收敛，通常为 30～50 秒。本章还介绍了减少 STP 收敛时间的措施：Uplinkfast、Backbonefast 和 RSTP 协议。默认时，Cisco 交换机为每个 VLAN 构建一棵树，这样方便控制 STP 树，但导致 STP 树数量太多。MST 则可以为多个 VLAN 共同构建一棵树。本章还介绍了保护 STP 树的几个简单措施：Root Guard、BPDU Guard、BPDU Filter 和环路保护。最后介绍了 Flexlink 技术，减少了链路发生故障时的切换时间。

第 4 章 VLAN 间路由

在交换机上划分 VLAN 后,不同 VLAN 间的计算机就无法通信了。VLAN 间的通信需要借助三层设备,我们可以使用路由器来实现这个功能,如果使用路由器通常会采用单臂路由模式。在实际工程中,VLAN 间的通信大多是通过三层交换机实现的,三层交换机可以看成路由器加交换机。然而因为思科的交换机采用了 CEF 技术,其数据处理能力比传统的、基于 CPU 的路由器要大得多。本章将分别介绍这两种方法的具体配置。实际上,现在的路由器也不采用基于过程的交换算法了,而是采用快速交换或者 CEF 交换算法,因此本章也将介绍路由器上不同的交换算法。

4.1 VLAN 间路由概述

4.1.1 使用路由器实现 VLAN 间的通信

用路由器实现 VLAN 间的通信如图 4-1 所示。两台 PC 虽然在同一交换机上,但是处于不同 VLAN,所以它们之间通信必须使用三层设备。要实现它们之间通信,可以在每个 VLAN 上都有一个以太网端口和路由器连接。然后在路由器的以太网端口配置 IP 地址(如果有必要可能还需要配置路由协议等),PC 上的网关指向同一 VLAN 中的路由器以太网端口上的 IP 即可。采用这种方法,如果要实现 N 个 VLAN 间的通信,则路由器需要 N 个以太网端口,同时也会占用交换机上的 N 个以太网端口,这在实际中并不可行。

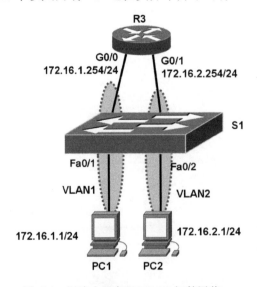

图 4-1 用路由器实现 VLAN 间的通信

4.1.2 单臂路由

如图 4-2 所示,单臂路由提供另外一种解决方案。路由器只需要一个以太网端口和交换机连接,交换机的这个端口被设置为 Trunk 端口。在路由器上创建多个子端口和不同的 VLAN 连接,子端口是路由器物理端口上的逻辑端口。工作原理如图 4-3 所示,当交换机收到 VLAN1 的计算机发送的数据帧后,从它的 Trunk 端口发送数据给路由器,由于该链路是 Trunk 链路,帧中带有 VLAN1 的标签,帧到达了路由器后,路由器在查询路由表后,如果数据要转发到 VLAN2 上,路由器将把数据帧重新用 VLAN2 的标签进行封装,通过 Trunk 链路发送到交换机上的 Trunk 端口;交换机收到该帧后,去掉 VLAN 2 标签,发送给 VLAN 2 上的计算机,从而实现了 VLAN 间的通信。Trunk 链路上同样也有 Native VLAN 问题,Native VLAN 的数据是不重新封装的。

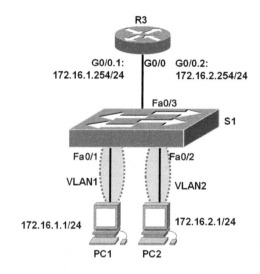

图 4-2　单臂路由实现 VLAN 间的通信

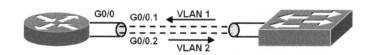

图 4-3　路由器的子端口工作原理

4.1.3 三层交换

采用单臂路由实现 VLAN 间的路由时转发速率较慢,而且需要昂贵的路由器设备。实际上,在局域网内部多采用三层交换机。三层交换机通常采用硬件来实现转发,其路由数据包的速率是普通路由器的几倍。

从使用者的角度可以把三层交换机看成二层交换机和路由器的组合,图 4-4 为三层交换机原理示意图。这个虚拟的路由器和每个 VLAN 都有一个端口进行连接,不过这个端口名称是

VLAN1 或 VLAN2 端口。在图 4-4 中，只要在 VLAN 端口上配置 IP 地址，PC 上的网关指向虚拟路由器的同名 VLAN 端口即可，即 VLAN 1 的 PC 网关是三层交换机上的 VLAN 1 端口的 IP 地址。在虚拟路由器上同样可以配置路由协议。思科的三层交换机是采用 CEF 技术来转发数据包的。

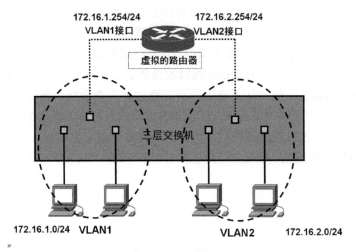

图 4-4　三层交换机原理示意图

4.1.4　路由器的三种交换算法

CEF 技术已经不仅仅在三层交换机上使用，路由器也使用 CEF 技术。因此路由器上有多种交换技术。

1. 过程交换

最初的 Cisco 路由器采用过程交换，路由器接收到数据包后，由 CPU 查找路由表后，再把数据包重新封装后从端口发送出去。这种方式的瓶颈在 CPU 上，很难满足大数据量的转发需求。

2. 快速交换（Fast Switching）

通常到达某特定目的地址的 IP 包是一个流，一个流不会只有一个数据包，而是多个数据包。当第一个数据包到达后，路由器先使用过程交换来进行转发，并且构建高速缓存。随后当同一个流的数据包到达时，路由器通过查找高速缓存进行转发，而不查找路由表。这样可以减少包在全路由表中查找同一目标的次数。这种"一次路由，多次交换"的方式称为快速交换，大大提高了路由器的包转发速率。IP 路由表的更新会使得高速缓存无效，在路由状况不断变化的环境中，路由高速缓存的优势将受到很大限制；同时现在的网络流量大多是发往 Internet 的，数据流很分散，将导致高速缓冲的急剧膨胀，维护和查找高速缓冲工作量也很大，因此快速交换也有很大的局限性。

快速交换（Fast Switching）通过生成并查找路由高速缓存交换数据包，该高速缓存的条目（包括目的 IP 地址、输出端口和 MAC 地址头信息等）是在第一个数据包到来时产生的，路由器对整个路由表执行最长匹配查找算法获得下一跳 IP 地址，然后查找 ARP 缓存获得二层的 MAC 地址信息并写入路由高速缓存，之后的数据包则根据已经生成的高速缓存的条目直接重写 MAC 头信息完成交换操作。

3. CEF 特快交换（CEF Switching）

CEF 是思科公司推出的高级三层交换技术，是为高性能、高伸缩性的三层 IP 骨干网交换设计的。作为优化包转发的路由查找机制，CEF 定义了两个主要部件：转发信息库（Forwarding Information Base, FIB）和邻接表（Adjacency Table）。

转发信息库（FIB）是路由器决定目标交换的查找表，FIB 的条目与 IP 路由表条目之间有一一对应的关系，即 FIB 是 IP 路由表中包含的路由信息的一个镜像。由于 FIB 包含了所有必需的路由信息，因此就不用再维护路由高速缓存了。当网络拓扑或路由发生变化时，IP 路由表被更新，FIB 的内容随之发生变化。

CEF 利用邻接表提供数据包 MAC 层重写所需的信息。FIB 中的每一项都指向邻接表里的某个下一跳中继段。若相邻节点间能通过数据链路层实现相互转发，则这些节点被列入邻接表中。系统一旦发现邻接关系，就将其写到邻接表中，邻接序列随时都在生成，每生成一个邻接条目，就会为那个邻接节点预先计算一个链路层头标信息，并把这个链路层头标信息存储在邻接表中，当决定路由时，它就指向下一网络段及相应的邻接条目。随后在对数据包进行 CEF 交换时，用它来进行封装。

欲查看邻接表的有关信息，与快速交换相似（见本节关于快速交换的介绍），CEF 也使用自己建立的数据结构（是 FIB 表，而不是路由表）来执行交换操作。但 CEF 通过 FIB 和邻接表对数据包进行交换，而 FIB 和邻接表是在数据包到来以前，由 CPU 根据路由表生成并定时更新的，因此到达路由器的第一个数据包也无须执行查找路由表的过程，直接由 FIB 和邻接表获得新的 MAC 头信息，就可进行交换了，即"一次都不路由，直接交换"。对于拥有大容量路由表的路由器来说，这种预先建立交换查找条目的方式能够有效地提高交换性能。由于 FIB 和邻接表的数量主要和路由表数量相关，而和流的数量无关，因此 CEF 更适合用于当今的数据转发（流很分散）。思科的三层交换机采用的是 CEF 技术。路由器则在不同的 IOS 版本中使用不同的交换技术，较新的 IOS 默认启用的是 CEF，较老的 IOS 默认启用的是快速交换。

4.2 实验1：采用单臂路由实现 VLAN 间路由

1. 实验目的

通过本实验可以掌握：
① 创建路由器以太网端口上的子端口的方法。
② 利用单臂路由完成 VLAN 间路由的配置。

2. 实验拓扑

实验拓扑如图 4-2 所示。注意：交换机 S1 实际上是三层交换机，但是这里并不启用其三层功能。

3. 实验步骤

我们用 R1 模拟 PC1，用 R2 模拟 PC2，用 R3 来实现分别位于 VLAN 1 和 VLAN 2 的 R1 和 R2 间的通信。S1 实际上是三层交换机，我们这里并不使用它的三层功能。

（1）在 S1 上配置 VLAN

```
S1(config)#vlan 2
S1(config-vlan)#exit
S1(config)#interface fastethernet0/1
S1(config-if)#switchport mode access
S1(config-if)#switchport access vlan 1
S1(config-if)#interface fastethernet0/2
S1(config-if)#switchport mode access
S1(config-if)#switchport access vlan 2
```

（2）先把交换机上的以太网端口配置成 Trunk 端口

```
S1(config)#interface fastethernet0/3
S1(config-if)#switch trunk encapsulation dot1q
S1(config-if)#switch mode trunk
```

（3）在路由器 R3 的物理以太网端口下创建子端口并定义封装的 VLAN

```
R3(config)#interface GigabitEthernet0/0
R3(config-if)#no shutdown
R3(config-if)#exit
R3(config)#interface GigabitEthernet0/0.1    //创建子端口，子端口的编号并无所谓
R3(config-subif)# encapsulation dot1q 1 native
```
//定义该子端口承载哪个 VLAN 流量，由于交换机上的 Native VLAN 是 VLAN1，所以这里也要指明该 VLAN 就是 Native VLAN。实际上，默认时 Native VLAN 就是 VLAN1
```
R1 (config-subif)#ip address 172.16.1.254 255.255.255.0
```
//在子端口上配置 IP 地址，这个地址就是 VLAN1 的网关
```
R3(config-if)#exit
R3(config)#interface GigabitEthernet 0/0.2
R3(config-subif)# encapsulation dot1q 2
R3 (config-subif)#ip address 172.16.2.254 255.255.255.0
```

（4）配置 R1 和 R2 模拟 PC

```
R1(config)#interface GigabitEthernet 0/0
R1(config-if)#no shutdown
R1(config-if)#ip address 172.16.1.1 255.255.255.0
R1(config-if)#exit
R1(config)#no ip routing    //关闭路由器的路由功能
R1(config)#ip default-gateway 172.16.1.254
```
//配置网关，和计算机上网络邻居配置网关类似，网关指向 R3 的 Fa0/0.1 子端口
```
R2(config)#interface GigabitEthernet 0/0
R2(config-if)#no shutdown
```

R2(config-if)#**ip address 172.16.2.1 255.255.255.0**
R2(config-if)#**exit**
R2(config)#**no ip routing**
R2(config)#**ip default-gateway 172.16.2.254**

4. 实验调试

（1）show ip route

```
R3#show ip route
Codes: L - local, C - connected, S - static, R - RIP, M - mobile, B - BGP
       D - EIGRP, EX - EIGRP external, O - OSPF, IA - OSPF inter area
       N1 - OSPF NSSA external type 1, N2 - OSPF NSSA external type 2
       E1 - OSPF external type 1, E2 - OSPF external type 2
       i - IS-IS, su - IS-IS summary, L1 - IS-IS level-1, L2 - IS-IS level-2
       ia - IS-IS inter area, * - candidate default, U - per-user static route
       o - ODR, P - periodic downloaded static route, H - NHRP, l - LISP
       a - application route
       + - replicated route, % - next hop override, p - overrides from PfR
Gateway of last resort is not set
       172.16.0.0/16 is variably subnetted, 4 subnets, 2 masks
C        172.16.1.0/24 is directly connected, GigabitEthernet0/0.1
L        172.16.1.254/32 is directly connected, GigabitEthernet0/0.1
C        172.16.2.0/24 is directly connected, GigabitEthernet0/0.2
L        172.16.2.254/32 is directly connected, GigabitEthernet0/0.2
```

以上输出表明，在 R3 上的路由表中可以看到两条直连路由。

（2）ping

```
R1#ping 172.16.2.1
Type escape sequence to abort.
Sending 5, 100-byte ICMP Echos to 172.16.2.1, timeout is 2 seconds:
!!!!!
Success rate is 100 percent (5/5), round-trip min/avg/max = 1/201/1000 ms
```

以上输出表明两个 VLAN 间已经可以通信了。

 提示

S1 实际上是 Catalyst 3560 交换机，该交换机具有三层功能，这里我们把它当作二层交换机使用了，有点大材小用。

4.3 实验2：采用三层交换实现 VLAN 间路由

1. 实验目的

通过本实验可以掌握：
① 三层交换的概念。
② 三层交换的配置。

2. 实验拓扑

采用三层交换实现 VLAN 间路由实验拓扑如图 4-5 所示。

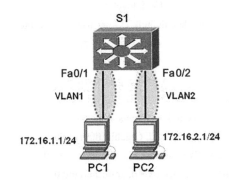

图 4-5 采用三层交换实现 VLAN 间路由实验拓扑

3. 实验步骤

用 R1 模拟 PC1，用 R2 模拟 PC2，用 S1 来实现分别位于 VLAN1 和 VLAN2 的 PC1 和 PC2 间的通信。

（1）在 S1 上划分 VLAN

```
S1(config)#vlan 2
S1(config-vlan)#exit
S1(config)#interface fastethernet0/1
S1(config-if)#switchport mode access
S1(config-if)#switchport access vlan 1
S1(config-if)#interface fastethernet0/2
S1(config-if)#switchport mode access
S1(config-if)#switchport access vlan 2
```

（2）配置三层交换

```
S1(config)#ip routing     //开启 S1 的路由功能，这时 S1 就启用了三层功能
S1(config)#int vlan 1
S1(config-if)#no shutdown
```

S1(config-if)#**ip address 172.16.1.254 255.255.255.0**
//在 VLAN 端口上配置 IP 地址即可，VLAN1 端口上的地址就是 PC1 的网关
S1(config-if)#**exit**
S1(config)#**int vlan 2**   //创建 VLAN2 端口，默认时，VLAN1 端口已经存在了
S1(config-if)#**no shutdown**   //如果本交换机某一 VLAN 有激活的端口存在，那么 VLAN 端口应该会自动 Up，我们这里还是执行了一次 no shutdown 命令
S1(config-if)#**ip address 172.16.2.254 255.255.255.0**
//在 VLAN 端口上配置 IP 地址，VLAN2 端口上的地址就是 PC2 的网关

 提示

要在三层交换机上启用路由功能，还需要启用 CEF（命令为 **ip cef**），不过这是默认值。和路由器一样，在三层交换机上同样可以运行路由协议。在三层交换机上可以有多个 VLAN 端口处于"Up"状态，这些端口被称为 SVI（Switch VLAN Interface），任何一个激活 SVI 都可以作为管理端口（即被 Telnet）。而在二层交换机上，虽然也可以有多个 VLAN 端口，但是只能有一个 VLAN 端口处于"Up"状态。

4．实验调试

（1）检查 IP 地址

S1#show ip interface brief			
Interface	IP-Address	OK? Method Status	Protocol
Vlan1	**172.16.1.254**	**YES manual up**	**up**
Vlan2	**172.16.2.254**	**YES manual up**	**up**
FastEthernet0/1	unassigned	YES unset up	up

（2）检查 S1 上的路由表

```
S1#show ip route
（此处省略）
     172.16.0.0/16 is variably subnetted, 4 subnets, 2 masks
C       172.16.1.0/24 is directly connected, Vlan1
L       172.16.1.254/32 is directly connected, Vlan1
C       172.16.2.0/24 is directly connected, Vlan2
L       172.16.2.254/32 is directly connected, Vlan2
```

和路由器一样，三层交换机上也有路由表。

（3）测试 VLAN 间的通信

① 配置 R1 和 R2 模拟 PC，如下所示：

R1(config)#**interface GigabitEthernet0/0**
R1(config-if)#**no shutdown**
R1(config-if)#**ip address 172.16.1.1 255.255.255.0**
R1(config)#**no ip routing**
R1(config)#**ip default-gateway 172.16.1.254**
R2(config)#**interface GigabitEthernet0/0**

```
R2(config-if)#no shutdown
R2(config-if)#ip address 172.16.2.1 255.255.255.0
R2(config)#no ip routing
R2(config)#ip default-gateway 172.16.2.254
```

② 测试 R1 和 R2 的连通性，如下所示：

```
R1#ping 172.16.2.1
Type escape sequence to abort.
Sending 5, 100-byte ICMP Echos to 172.16.2.1, timeout is 2 seconds:
!!!!!
Success rate is 100 percent (5/5), round-trip min/avg/max = 1/2/4 ms
```

从以上输出可以看到，VLAN 间已经可以通信了。

提示

三层交换机可以实现 VLAN 间的数据通信，但是并不要求这些计算机一定接在三层交换机上。如图 4-6 所示，PC1 和 PC2 并不直接接在交换机 S1 上，它们之间的通信同样可以通过 S1 实现。图中交换机 S1 除了要完成本节之前介绍的配置，还需要保证 S1 和 S2 之间的 Trunk 链路是正常的。PC1 发往 PC2 的数据，首先到 S2，经过 Trunk 链路（Fa0/23 端口）到达 S1，S1 进行三层交换，数据经过 Trunk 链路（Fa0/23 端口）到达 S2，S2 发送给 PC2。这期间二层的帧和三层 IP 包均会发生改变。

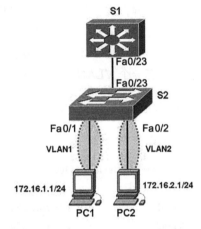

图 4-6 计算机不是直接接在三层交换机上

提示

还有另一种方法也可以实现 VLAN 间的通信，其拓扑如图 4-5 所示，具体配置如下：

```
S1(config)#ip routing      //在交换机上启用路由功能
S1(config)#interface fastethernet0/1
S1(config-if)#no switchport
```
//使得 Fa0/1 不再是交换端口了，成为了路由端口（三层端口），Fa0/1 和一般路由器的 Fa0/1 没有什么差别了
```
S1(config-if)#ip address 172.16.1.254 255.255.255.0
S1(config)#interface fastethernet0/2
S1(config-if)#no switchport
S1(config-if)#ip address 172.16.2.254 255.255.255.0
```

计算机的配置和本节中的配置是一样的，这样 VLAN 1 和 VLAN 2 中的计算机就可以互相通信了，采用这种方法实际上把三层交换机当成路由使用了。如果 S1 上的全部以太网都这样设置，S1 实际上成了具有 24 个以太网端口的路由器了，我们不建议这样做，这样太浪费端口了。但是在核心层交换机上，常常采用三层端口和分布层交换机连接，这样可以避免在核心层交换机上运行 STP 等二层协议。

4.4 实验 3：在三层交换机上配置路由协议

1. 实验目的

通过本实验可以掌握：
① 三层端口的概念。
② 在三层交换机上配置路由协议的方法。

2. 实验拓扑

在三层交换机上配置路由协议实验拓扑如图 4-7 所示。交换机 S1 和 S2 之间采用链路汇聚，交换机之间的链路使用三层端口，在交换机上运行 EIGRP。

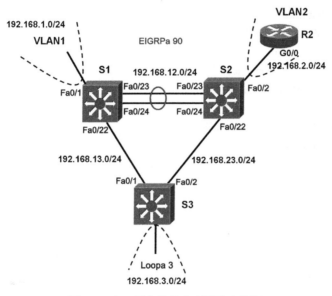

图 4-7 在三层交换机上配置路由协议

3. 实验步骤

（1）在交换机上创建 VLAN

```
S2(config)#vlan 2
S2(config-vlan)#exit
```

（2）启用三层端口，配置 IP 地址

```
S1(config)#interface port-channel 1
S1(config-if)#no switchport
//把链路汇聚端口改为三层端口
S1(config-if)#ip address 192.168.12.1 255.255.255.0
```

```
S1(config)#interface fastEthernet 0/22
S1(config-if)#no switchport
S1(config-if)#ip address 192.168.13.1 255.255.255.0
S1(config)#int vlan 1
S1(config-if)#no shutdown
S1(config-if)#ip address 192.168.1.1 255.255.255.0

S2(config)#interface port-channel 1
S2(config-if)#no switchport
S2(config-if)#ip address 192.168.12.2 255.255.255.0
S2(config)# interface fastEthernet 0/22
S2(config-if)#no switchport
S2(config-if)#ip address 192.168.23.2 255.255.255.0
S2(config)#int vlan 2
S2(config-if)#no shutdown
S2(config-if)#ip address 192.168.2.2 255.255.255.0

S2(config)# interface fastEthernet 0/2
S2(config-if)#switchport mode access
S2(config-if)#switchport access vlan 2
R2(config)# interface GigabitEthernet 0/2
R2(config-if)#no shutdown
// 以上 5 条配置命令是为了使交换机 S2 的 VLAN 2 端口处于 Up 状态

S3(config)# interface fastEthernet 0/1
S3(config-if)#no switchport
S3(config-if)#ip address 192.168.13.3 255.255.255.0
S3(config)# interface fastEthernet 0/2
S3(config-if)#no switchport
S3(config-if)#ip address 192.168.23.3 255.255.255.0
S3(config)#int loopback3
S3(config-if)#ip address 192.168.3.3 255.255.255.0
```

> **提示**
>
> 交换机的 VLAN 端口处于 Up 状态的条件：至少有属于该 VLAN 的端口处于 Up 状态或者有 Trunk 端口处于 Up 状态。因此，为保证 S1 和 S2 的 VLAN 端口处于 Up 状态，可以把一些端口划分到相应的 VLAN，并且把端口上的设备接入。

（3）创建链路汇聚

```
S1(config)#interface range f0/23 -24
S1(config-if-range)# no switchport
S1(config-if-range)#channel-group 1 mode on

S2(config)#interface range f0/23 -24
```

```
S2(config-if-range)# no switchport
S2(config-if-range)#channel-group 1 mode on
```

(4)配置路由协议

```
S1(config)#ip routing
//启用三层交换功能
S1(config)#router eigrp 90
//启用路由协议,不同 IOS 版本可能支持不同的协议。交换机上的路由协议配置和路由器上是一样的
S1(config-router)#network 192.168.1.0
S1(config-router)#network 192.168.12.0
S1(config-router)#network 192.168.13.0

S2(config)#ip routing
S2(config)#router eigrp 90
S2(config-router)#network 192.168.2.0
S2(config-router)#network 192.168.12.0
S2(config-router)#network 192.168.23.0

S3(config)#ip routing
S3(config)#router eigrp 90
S3(config-router)#network 192.168.3.0
S3(config-router)#network 192.168.13.0
S3(config-router)#network 192.168.23.0
```

4. 实验调试

(1)查看路由表

```
S1#show ip route
         192.168.1.0/24 is variably subnetted, 2 subnets, 2 masks
C           192.168.1.0/24 is directly connected, Vlan1
L           192.168.1.1/32 is directly connected, Vlan1
D        192.168.2.0/24 [90/15616] via 192.168.12.2, 00:00:14, Port-channel1
D        192.168.3.0/24 [90/156160] via 192.168.13.3, 00:04:21, FastEthernet0/22
         192.168.12.0/24 is variably subnetted, 2 subnets, 2 masks
C           192.168.12.0/24 is directly connected, Port-channel1
L           192.168.12.1/32 is directly connected, Port-channel1
         192.168.13.0/24 is variably subnetted, 2 subnets, 2 masks
C           192.168.13.0/24 is directly connected, FastEthernet0/22
L           192.168.13.1/32 is directly connected, FastEthernet0/22
D        192.168.23.0/24 [90/30720] via 192.168.13.3, 00:04:21, FastEthernet0/22
                        [90/30720] via 192.168.12.2, 00:04:21, Port-channel1


S2#show ip route
D        192.168.1.0/24 [90/15616] via 192.168.12.1, 00:12:04, Port-channel1
```

```
            192.168.2.0/24 is variably subnetted, 2 subnets, 2 masks
C           192.168.2.0/24 is directly connected, Vlan2
L           192.168.2.2/32 is directly connected, Vlan2
D           192.168.3.0/24 [90/156160] via 192.168.23.3, 00:12:09, FastEthernet0/22
            192.168.12.0/24 is variably subnetted, 2 subnets, 2 masks
C           192.168.12.0/24 is directly connected, Port-channel1
L           192.168.12.2/32 is directly connected, Port-channel1
D           192.168.13.0/24 [90/30720] via 192.168.23.3, 00:12:09, FastEthernet0/22
                           [90/30720] via 192.168.12.1, 00:12:09, Port-channel1
            192.168.23.0/24 is variably subnetted, 2 subnets, 2 masks
C           192.168.23.0/24 is directly connected, FastEthernet0/22
L           192.168.23.2/32 is directly connected, FastEthernet0/22

S3#show ip route
D           192.168.1.0/24 [90/28416] via 192.168.13.1, 00:12:28, FastEthernet0/1
D           192.168.2.0/24 [90/28416] via 192.168.23.2, 00:08:20, FastEthernet0/2
            192.168.3.0/24 is variably subnetted, 2 subnets, 2 masks
C           192.168.3.0/24 is directly connected, Loopback3
L           192.168.3.3/32 is directly connected, Loopback3
D           192.168.12.0/24 [90/30720] via 192.168.23.2, 00:12:28, FastEthernet0/2
                            [90/30720] via 192.168.13.1, 00:12:28, FastEthernet0/1
            192.168.13.0/24 is variably subnetted, 2 subnets, 2 masks
C           192.168.13.0/24 is directly connected, FastEthernet0/1
L           192.168.13.3/32 is directly connected, FastEthernet0/1
            192.168.23.0/24 is variably subnetted, 2 subnets, 2 masks
C           192.168.23.0/24 is directly connected, FastEthernet0/2
L           192.168.23.3/32 is directly connected, FastEthernet0/2
```

（2）查看 CEF 表

```
S1#show ip cef
//以下显示 FIB 表
Prefix              Next Hop          Interface
0.0.0.0/0           no route
0.0.0.0/8           drop
0.0.0.0/32          receive
127.0.0.0/8         drop
192.168.1.0/24      attached          Vlan1
//attached 表示是直连链路
192.168.1.0/32      receive           Vlan1
192.168.1.1/32      receive           Vlan1
                    //receive 表示数据将交给 CPU 查找路由表，而不是 CEF 交换
192.168.1.255/32    receive           Vlan1
192.168.2.0/24      192.168.12.2      Port-channel1
//指明下一跳是 192.168.12.2，在邻接表有该下一跳的二层信息，因此数据包会直接转发
```

192.168.3.0/24	192.168.13.3	FastEthernet0/22
192.168.12.0/24	attached	Port-channel1
192.168.12.0/32	receive	Port-channel1
192.168.12.1/32	receive	Port-channel1
192.168.12.2/32	attached	Port-channel1
192.168.12.255/32	receive	Port-channel1

（略去部分输出）

S1#**show ip cef detail**
IPv4 CEF is enabled for distributed and running   //采用分布式的 CEF
VRF Default
　25 prefixes (25/0 fwd/non-fwd)
　Table id 0x0
　Database epoch:　　　　3 (25 entries at this epoch)

0.0.0.0/0, epoch 3, flags default route handler, default route
　no route
0.0.0.0/8, epoch 3
　Special source: drop
　drop
0.0.0.0/32, epoch 3, flags receive
　Special source: receive
　receive
127.0.0.0/8, epoch 3
　Special source: drop
　drop
192.168.1.0/24, epoch 3, flags attached, connected, cover dependents, need deagg
　Interest List:
　　- ipv4fib connected receive
　Covered dependent prefixes: 2
　　need deagg: 2
　attached to Vlan1
192.168.1.0/32, epoch 3, flags receive
　Interface source: Vlan1
　Dependent covered prefix type cover need deagg, cover 192.168.1.0/24
　receive for Vlan1
192.168.1.1/32, epoch 3, flags receive, source eligible
　Interface source: Vlan1
　receive for Vlan1
192.168.1.255/32, epoch 3, flags receive
　Interface source: Vlan1
　Dependent covered prefix type cover need deagg, cover 192.168.1.0/24
　receive for Vlan1
192.168.2.0/24, epoch 3

```
      nexthop 192.168.12.2 Port-channel1
  192.168.3.0/24, epoch 3
      nexthop 192.168.13.3 FastEthernet0/22
  192.168.12.0/24, epoch 3, flags attached, connected, cover dependents, need deagg
      Interest List:
          - ipv4fib connected receive
      Covered dependent prefixes: 3
          need deagg: 2
          notify cover updated: 1
      attached to Port-channel1
  192.168.12.0/32, epoch 3, flags receive
      Interface source: Port-channel1
      Dependent covered prefix type cover need deagg, cover 192.168.12.0/24
      receive for Port-channel1
  192.168.12.1/32, epoch 3, flags receive, source eligible
      Interface source: Port-channel1
      receive for Port-channel1
  192.168.12.2/32, epoch 3, flags attached
      Adj source: IP adj out of Port-channel1, addr 192.168.12.2 05977A40
      Dependent covered prefix type adjfib, cover 192.168.12.0/24
      attached to Port-channel1
(略去部分输出)
```

（3）查看邻接表

```
S1#show adjacency
Protocol Interface                 Address
IP       FastEthernet0/22          192.168.13.3(22)
IP       Port-channel1             192.168.12.2(22)
//以上是 S1 的邻接表

S1#show adjacency detail
//以下是详细的邻接表信息
Protocol Interface                 Address
IP       FastEthernet0/22          192.168.13.3(22)
                                   0 packets, 0 bytes
                                   epoch 0
                                   sourced in sev-epoch 0
                                   Encap length 14
                                   D0C789C283C1D0C789AB11C20800
//重写的二层信息，前 16 位是源 MAC 地址，紧接的 16 位是目的 MAC 地址，0800 表示是 IP 包
                                   L2 destination address byte offset 0
                                   L2 destination address byte length 6
                                   Link-type after encap: ip
                                   ARP
```

第 4 章 VLAN 间路由

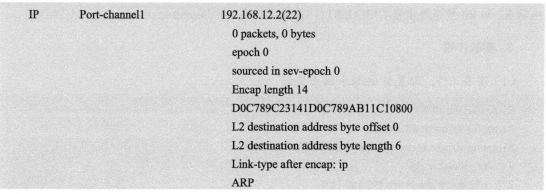

IP	Port-channel1	192.168.12.2(22)
		0 packets, 0 bytes
		epoch 0
		sourced in sev-epoch 0
		Encap length 14
		D0C789C23141D0C789AB11C10800
		L2 destination address byte offset 0
		L2 destination address byte length 6
		Link-type after encap: ip
		ARP

> 📁 **提示**
>
> 基于 CEF 的排错方法：
>
> ① 使用 **show ip route** 命令查看路由表，如果路由表不存在或者出现下一跳错误，则修改路由协议的配置。
>
> ② 使用 **show ip arp** 命令查看下一跳 IP 地址对应的 MAC 是否正确，如果不正确或者条目不存在，则对 ARP 进行排错（例如，清除 arp 表）。
>
> ③ 使用 **show ip cef** 查看 CEF 表，检查 FIB 表的下一跳与路由表的一下跳是否一致。
>
> ④ 使用 **show adjacency detail** 命令查看邻接表中的二层重写信息是否与 ARP 一致。

4.5 实验 4：路由器上的 3 种交换方式

1. 实验目的

通过本实验可以掌握：
① 基于过程的交换方式。
② 快速交换（Fast Switching）方式。
③ CEF 交换（CEF Switching）方式。

2. 实验拓扑

3 种交换方式实验拓扑如图 4-8 所示。

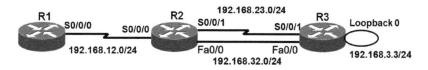

图 4-8 3 种交换方式实验拓扑

> 📁 **提示**
>
> 不同的 IOS 使用的交换方式差异很大，笔者发现，在 "c2900-universalk9-mz.SPA.157-3.M.bin" IOS 中，即使配置是正常的，也无法使用快速交换方式。因此，本实验中将路

由器 R2 和 R3 换成路由器 CISCO2811，并且 IOS 是 "c2800nm-ipbase-mz.123-14.T4.bin"。

3. 实验步骤

（1）准备工作：配置 IP 地址和路由协议

```
R1(config)#int s0/0/0
R1(config-if)#no shutdown
R1(config-if)#ip address 192.168.12.1 255.255.255.0
R1(config)#router rip
R1(config-router)#network 192.168.12.0

R2(config)#int s0/0/0
R2(config-if)#no shutdown
R2(config-if)#clock rate 128000
R2(config-if)#ip address 192.168.12.2 255.255.255.0
R2(config)#int s0/0/1
R2(config-if)#no shutdown
R2(config-if)#clock rate 128000
R2(config-if)#ip address 192.168.23.2 255.255.255.0
R2(config)#int f0/0
R2(config-if)#no shutdown
R2(config-if)#ip address 192.168.32.2 255.255.255.0
R2(config)#router rip
R2(config-router)#network 192.168.12.0
R2(config-router)#network 192.168.32.0
R2(config-router)#network 192.168.23.0

R3(config)#int loopback 0
R3(config-if)#ip address 192.168.3.3 255.255.255.0
R3(config)#int s0/0/1
R3(config-if)#no shutdown
R3(config-if)#ip address 192.168.23.3 255.255.255.0
R3(config)#int f0/0
R2(config-if)#no shutdown
R3(config-if)#ip address 192.168.32.3 255.255.255.0
R3(config)#router rip
R3(config-router)#network 192.168.3.0
R3(config-router)#network 192.168.32.0
R3(config-router)#network 192.168.23.0
```

（2）在 R3 上打开 debug

```
R3(config)#access-list 101 permit icmp host 192.168.12.1 any
//定义 ACL，只显示从 R1 发送过来的 ICMP（ping 命令）数据包
R3#debug ip packet 101
```

IP packet debugging is on for access list 101
//打开 debug，将动态显示收到的数据包情况

（3）CEF 交换方式

这里的路由器默认是不启用 CEF 的，但快速交换是开启的。开启 CEF 后，路由器会优先采用 CEF。不同 IOS 版本的情况不同，有的默认启用 CEF，有的默认只启用快速转发方式并不启用 CEF。

R2(config)#**ip cef**
//启用 CEF
R2#show ip interface s0/0/0
Serial0/0/0 is up, line protocol is up
　Internet address is 192.168.12.2/24
　（略去部分输出）
　ICMP unreachables are always sent
　ICMP mask replies are never sent
　IP fast switching is enabled　　//快速交换是启用的
　IP fast switching on the same interface is enabled
　IP Flow switching is disabled
　IP CEF switching is enabled　　//CEF 交换是启用的
　IP CEF switching turbo vector
　IP multicast fast switching is enabled
　IP multicast distributed fast switching is disabled
　IP route-cache flags are Fast, CEF
//高速缓冲标志是快速交换和 CEF，如果快速交换和 CEF 是同时开启的，则路由器会使用 CEF 交换方式
　Router Discovery is disabled
　IP output packet accounting is disabled
　（略去部分输出）

R2#show ip interface f0/0
　（略去部分输出）
　ICMP redirects are always sent
　ICMP unreachables are always sent
　ICMP mask replies are never sent
　IP fast switching is enabled
　IP fast switching on the same interface is disabled
　IP Flow switching is disabled
　IP CEF switching is enabled
　IP CEF switching turbo vector
　IP multicast fast switching is enabled
　IP multicast distributed fast switching is disabled
　IP route-cache flags are Fast, CEF
　（略去部分输出）

```
R2#show ip cef
//有 FIB 表，以下进一步证明路由器开启了 CEF
Prefix              Next Hop            Interface
0.0.0.0/0           no route
0.0.0.0/8           drop
0.0.0.0/32          receive
127.0.0.0/8         drop
192.168.3.0/24      192.168.23.3        Serial0/0/1
                    192.168.32.3        FastEthernet0/0
192.168.12.0/24     attached            Serial0/0/0
192.168.12.0/32     receive             Serial0/0/0
192.168.12.2/32     receive             Serial0/0/0
192.168.12.255/32   receive             Serial0/0/0
192.168.23.0/24     attached            Serial0/0/1
192.168.23.0/32     receive             Serial0/0/1
192.168.23.2/32     receive             Serial0/0/1
（略去部分输出）
```

从 R1 上 ping 192.168.3.3，则在 R3 上可以看到：

```
*Mar   1 00:43:19.328: IP: s=192.168.12.1 (FastEthernet0/0), d=192.168.3.3, len 100, rcvd 4
*Mar   1 00:43:19.344: IP: s=192.168.12.1 (FastEthernet0/0), d=192.168.3.3, len 100, rcvd 4
*Mar   1 00:43:19.368: IP: s=192.168.12.1 (FastEthernet0/0), d=192.168.3.3, len 100, rcvd 4
*Mar   1 00:43:19.384: IP: s=192.168.12.1 (FastEthernet0/0), d=192.168.3.3, len 100, rcvd 4
*Mar   1 00:43:19.408: IP: s=192.168.12.1 (FastEthernet0/0), d=192.168.3.3, len 100, rcvd 4
```

从以上的输出可以看到，R3 从 Fa0/0 端口收到了 R1 发送的 5 个数据包，也就是说 R2 并没有使用负载均衡来发送数据包。这是因为 R2 使用 CEF 交换方式，而 CEF 默认是基于目的 IP 的负载均衡（即到达同一目的 IP 的数据包将使用同一链路发送）。

```
R2#show ip cache
IP routing cache 0 entries, 0 bytes
    0 adds, 0 invalidates, 0 refcounts
Minimum invalidation interval 2 seconds, maximum interval 5 seconds,
    quiet interval 3 seconds, threshold 0 requests
Invalidation rate 0 in last second, 0 in last 3 seconds

Prefix/Length           Age         Interface           Next Hop
        //由于 R2 现在使用的是 CEF，不是快速交换，因此高速缓冲表是空的
```

以下改变 R2 上的 CEF 的负载均衡方式为每数据包均衡：

```
R2(config)#interface FastEthernet0/0
R2(config-if)#ip load-sharing per-packet
R2(config)#interface Serial0/0/1
R2(config-if)#ip load-sharing per-packet
```

从 R1 上 ping 192.168.3.3，则在 R3 上可以看到：

```
*Mar   1 00:44:45.361: IP: s=192.168.12.1 (Serial0/0/1), d=192.168.3.3, len 100, rcvd 4
```

```
*Mar   1 00:44:45.377: IP: s=192.168.12.1 (FastEthernet0/0), d=192.168.3.3, len 100, rcvd 4
*Mar   1 00:44:45.409: IP: s=192.168.12.1 (Serial0/0/1), d=192.168.3.3, len 100, rcvd 4
*Mar   1 00:44:45.425: IP: s=192.168.12.1 (FastEthernet0/0), d=192.168.3.3, len 100, rcvd 4
*Mar   1 00:44:45.453: IP: s=192.168.12.1 (Serial0/0/1), d=192.168.3.3, len 100, rcvd 4
```

从以上的输出可以看到，R3 从 Fa0/0 和 S0/0/1 端口轮流收到了 R1 发送的 5 个数据包，也就是说，R2 已经使用负载均衡来发送数据包了。

> **提示**
>
> 配置负载均衡应该在所有到达目的 IP 的出端口上进行，建议清除一下路由表。

（4）快速交换方式

```
R2(config)#no ip cef
//在全局模式下关闭了 CEF 功能

R2#show ip interface s0/0/1
Serial0/0/1 is up, line protocol is up
  Internet address is 192.168.23.2/24
  （略去部分输出）
  IP fast switching is enabled     //快速转发是开启的
  IP fast switching on the same interface is disabled
  IP Flow switching is disabled
  IP CEF switching is disabled     //CEF 已经关闭了
  IP Fast switching turbo vector
  IP multicast fast switching is enabled
  IP multicast distributed fast switching is disabled
  IP route-cache flags are Fast    //高速缓冲的标志仅为了进行快速交换，即 CEF 交换被关闭
  （略去部分输出）

R2#show ip interface f0/0
FastEthernet0/0 is up, line protocol is up
  Internet address is 192.168.32.2/24
  （略去部分输出）
  IP fast switching is enabled
  IP fast switching on the same interface is disabled
  IP Flow switching is disabled
  IP CEF switching is disabled
  IP Fast switching turbo vector
  IP multicast fast switching is enabled
  IP multicast distributed fast switching is disabled
  IP route-cache flags are Fast
  （略去部分输出）
```

从 R1 上 ping 192.168.3.3，则在 R3 上可以看到：

```
*Mar   1 00:48:43.315: IP: s=192.168.12.1 (FastEthernet0/0), d=192.168.3.3, len 100, rcvd 4
*Mar   1 00:48:43.339: IP: s=192.168.12.1 (FastEthernet0/0), d=192.168.3.3, len 100, rcvd 4
*Mar   1 00:48:43.355: IP: s=192.168.12.1 (FastEthernet0/0), d=192.168.3.3, len 100, rcvd 4
```

```
*Mar  1 00:48:43.379: IP: s=192.168.12.1 (FastEthernet0/0), d=192.168.3.3, len 100, rcvd 4
*Mar  1 00:48:43.395: IP: s=192.168.12.1 (FastEthernet0/0), d=192.168.3.3, len 100, rcvd 4
```
//由于使用了快速交换，R3 从同一端口接收数据包

```
R2#show ip cache
IP routing cache 2 entries, 332 bytes
    14 adds, 12 invalidates, 0 refcounts
Minimum invalidation interval 2 seconds, maximum interval 5 seconds,
    quiet interval 3 seconds, threshold 0 requests
Invalidation rate 0 in last second, 0 in last 3 seconds
Last full cache invalidation occurred 00:00:29 ago
Prefix/Length            Age          Interface          Next Hop
192.168.3.3/32           00:00:25     FastEthernet0/0    192.168.32.3
192.168.12.1/32          00:00:25     Serial0/0          192.168.12.1
```
//可以看到高速缓冲，证明路由器使用的是快速交换

技术要点

可以使用两种方式来关闭 CEF 功能：

① 在全局模式下使用 "**no ip cef**" 命令。

② 在全局模式下保持 **ip cef**，在端口上使用 "**no ip route-cache cef**" 命令单独关闭端口的 CEF 功能。

（5）过程交换方式

```
R2(config)#no ip cef
```
//在全局模式下关闭了 CEF 功能
```
R2(config-if)#int s0/0/1
R2(config-if)#no ip route-cache
```
//以上命令关闭快速交换。注意该命令和 "**no ip route-cache cef**" 命令的不同
```
R2(config-if)#int f0/0
R2(config-if)#no ip route-cache

R2#show ip interface
FastEthernet0/0 is up, line protocol is up
    Internet address is 192.168.32.2/24
        （略去部分输出）
    IP fast switching is disabled           //快速交换功能关闭了
    IP fast switching on the same interface is disabled
    IP Flow switching is disabled
    IP CEF switching is disabled
    IP Fast switching turbo vector
    IP multicast fast switching is disabled
    IP multicast distributed fast switching is disabled
    IP route-cache flags are None           //没有了快速缓冲
```

（略去部分输出）
Serial0/0/1 is up, line protocol is up
  Internet address is 192.168.23.2/24
   （略去部分输出）
  IP fast switching is disabled   //快速交换功能关闭了
  IP fast switching on the same interface is disabled
  IP Flow switching is disabled
  IP CEF switching is disabled
  IP Fast switching turbo vector
  IP multicast fast switching is disabled
  IP multicast distributed fast switching is disabled
  IP route-cache flags are None   //没有快速缓冲
   （略去部分输出）
R2#**clear ip cache**
//清除快速缓冲表

从 R1 ping 192.168.3.3，则 R3 上可以看到：
 *Mar 1 00:53:45.695: IP: s=192.168.12.1 **(Serial0/0/1)**, d=192.168.3.3, len 100, rcvd 4
 *Mar 1 00:53:45.719: IP: s=192.168.12.1 **(FastEthernet0/0)**, d=192.168.3.3, len 100, rcvd 4
 *Mar 1 00:53:45.743: IP: s=192.168.12.1 **(Serial0/0/1)**, d=192.168.3.3, len 100, rcvd 4
 *Mar 1 00:53:45.767: IP: s=192.168.12.1 **(FastEthernet0/0)**, d=192.168.3.3, len 100, rcvd 4
 *Mar 1 00:53:45.791: IP: s=192.168.12.1 **(Serial0/0/1)**, d=192.168.3.3, len 100, rcvd 4
//由于 R2 不采用快速交换方式，使用过程交换方式，因此 R2 会进行负载均衡，R3 会从不同的端口收到数据包

R2#**show ip cache**
IP routing cache 1 entry, 160 bytes
    16 adds, 15 invalidates, 0 refcounts
Minimum invalidation interval 2 seconds, maximum interval 5 seconds,
    quiet interval 3 seconds, threshold 0 requests
Invalidation rate 0 in last second, 0 in last 3 seconds
Last full cache invalidation occurred 00:01:04 ago
Prefix/Length   Age   Interface   Next Hop
192.168.12.1/32   00:00:01 Serial0/0/0   192.168.12.1
//没有看到 192.168.3.3 的缓冲条目，然而由于 S0/0/0 端口上快速缓冲功能还开启着，因此有到 192.168.12.1 的缓冲条目

技术要点

如果在路由上配置了各种交换方法，确定最终交换方法如表 4-1 所示。总结如下：
① 如果入端口是 CEF 方式，则无论出端口是什么方式，都采用 CEF 方式。
② 如果入端口没有 CEF 方式，而出端口采用 CEF 方式，则采用快速方式。
③ 如果入端口没有 CEF 方式，而出端口也没有采用 CEF 方式，则采用出端口的设置方式。

表 4-1 路由器交换方法的确认

入端口配置	出端口配置	最 终 结 果
CEF	过程	CEF
CEF	快速	CEF
CEF	CEF	CEF
快速	过程	过程
快速	快速	快速
快速	CEF	CEF
过程	过程	过程
过程	快速	快速
过程	CEF	CEF

此外，快速交换对路由器自己产生的数据包不起作用。

4.6 本 章 小 结

本章介绍了实现不同 VLAN 间的计算机通信方法。可以使用单臂路由方法，在路由器的以太网端口上创建子端口。然而通常采用三层交换机来实现 VLAN 间的路由，三层交换机可以看成交换机和路由器的集成，配置三层交换机非常简单。本章最后介绍了思科的 3 种交换方式：过程交换方式、快速交换方式和 CEF 交换方式。

第 5 章 高 可 用 性

为了减少交换机故障对网络的影响，交换机采用了 STP 技术。然而作为网关的三层交换机或者路由器出故障了，又有什么解决办法呢？HSRP 和 VRRP 是最常用的网关冗余技术，HSRP 和 VRRP 类似，由多个三层交换机或者路由器共同组成一个组，虚拟出一个网关，其中的一台路由器处于活动状态，当它出故障时由备份路由器接替它的工作，从而实现对用户的透明切换。然而我们希望在实现冗余的同时，能同时实现负载平衡，以充分利用设备的能力，GLBP 同时提供了冗余和负载平衡能力。除了网关的负载平衡，服务器也需要负载平衡，思科的某些型号设备也能实现服务器负载均衡。对网络设备的监控也是很重要的工作，网络设备的日志和重要事件可以发送到服务器上，供管理员监控，以提高网络的可用性。本章将介绍它们的具体配置。

5.1 高可用性技术简介

5.1.1 HSRP

HSRP（Hot Standby Router Protocol）是 Cisco 的专有协议。HSRP 把多台路由器组成一个"热备份组"，形成一个虚拟路由器。这个组内只有一个路由器是活动的（Active），并由它来转发数据包，如果活动路由器发生故障，备份路由器将成为新的活动路由器。从网络内的主机来看，网关并没有改变。

HSRP 路由器利用 Hello 包来互相监听各自的存在。当路由器长时间没有接收到 Hello 包时，就认为活动路由器发生故障，备份路由器就会成为活动路由器。HSRP 协议利用优先级决定哪个路由器成为活动路由器。如果一个路由器的优先级比其他路由器的优先级高，则该路由器成为活动路由器。路由器的默认优先级是 100。在一个组中，最多有一台活动路由器和一台备份路由器。

HSRP 路由器发送的组播（224.0.0.2）消息有以下 3 种。

① Hello：Hello 消息通知其他路由器发送者的 HSRP 优先级和状态信息，HSRP 路由器默认为每 3 秒发送一个 Hello 消息。

② Coup：当备用路由器变为活动路由器时发送一个 Coup 消息。

③ Resign：当活动路由器宕机或者当有优先级更高的路由器发送 Hello 消息时，主动发送一个 Resign 消息。

HSRP 协议数据包格式如图 5-1 所示。

0 7	8 15	16 23	24 31
版本	操作码	状态	呼叫时间
等待时间	优先级	组	保留
保留			
认证码			
虚拟IP			

图 5-1 HSRP 协议数据包格式

① 版本：指示 HSPR 的版本信息。

② 操作码：用来描述数据包中报文的类型，可能的值为 0、1 和 2，分别表示是 Hello、Coup 和 Resign 消息。

③ 状态：描述发出该报文的路由器的当前状态。有 0、1、2、4、8 和 16 六种状态，分别表示的是 Initial、Learn、Listen、Speak、Standby 和 Active 状态。

④ 呼叫时间（Hellotime）：只在呼叫报文中有意义，表示路由器定时发送呼叫报文的间隔时间，以秒为单位。如果该参数没有在路由器上配置，它可以从活动路由器上学习获得。默认值为 3 秒。

⑤ 保持时间（Holdtime）：只在呼叫报文中有意义，被接收路由器用来判断该呼叫报文是否合法，单位为秒，其值至少是呼叫时间的 3 倍。如果该参数没有配置，也同样可以从活动路由器上学习。活动路由器不能从等待路由器上学习呼叫时间和保持时间，它只能继续使用从先前的活动路由器上学习来的该值。默认值为 10 秒。

⑥ 优先级：该参数用来选择活动路由器和等待路由器，两个具有不同优先级的路由器，优先级高的将成为活动路由器。两个具有相同优先级的路由器，IP 地址大的将成为活动路由器。默认优先级为 100。

⑦ 组：用来标记路由器所在的热等待组。对令牌环类型的网络，合法的值是 0、1 和 2，对于其他类型的网络，合法值是 0~255。

⑧ 认证码：包括 8 个明文的字符作为密码，如果没有配置，默认值为 cisco 。

⑨ 虚拟 IP 地址：用来指定本热等待组的虚拟 IP 地址，它可以从活动路由器的呼叫报文中学习到。如果没有配置该地址，并且呼叫报文是需要认识的，那么只能通过活动路由器学习。

HSRP 路由器有以下 6 种状态。

① Initial：HSRP 启动时的状态，HSRP 还没有运行，一般是在改变配置或端口刚刚启动时进入该状态。

② Learn：在该状态下，路由器还没有决定虚拟 IP 地址，也没有看到认证的、来自活动路由器的 Hello 报文。路由器仍在等待活动路由器发来的 Hello 报文。

③ Listen：路由器已经得到了虚拟 IP 地址，但是它既不是活动路由器也不是等待路由器。它一直监听从活动路由器和等待路由器发来的 Hello 报文。

④ Speak：在该状态下，路由器定期发送 Hello 报文，并且积极参加活动路由器或等待路由器的竞选。如果选出活动路由和等待路由，则变成监听状态（Listen）。而此时只有活动路由和等待路由处于说话状态。

⑤ Standby：处于该状态的路由器是下一个候选的活动路由器，它定时发送 Hello 报文。

⑥ Active：处于活动状态的路由器承担转发数据包的任务，这些数据包是发给该组的虚拟 MAC 地址的。它定时发出 Hello 报文。

HSRP 使用两个计时器：Hello 间隔和 Hold 间隔。默认的 Hello 间隔是 3 秒，默认的 Hold 间隔是 10 秒。Hello 间隔定义了同组路由器之间交换消息的频率。Hold 间隔定义了经过多长时间后，活跃路由器或者备用路由器就会被宣告为失败。配置计时器并不是越小越好，虽然计时器越小则切换时间越短。计时器的配置需要和 STP 等的切换时间相一致。另外，Hold 间隔最少应该是 Hello 间隔的 3 倍。

5.1.2 VRRP

VRRP 的工作原理和 HSRP 的非常类似，不过 VRRP 是国际标准，允许在不同厂商的设备之间运行。VRRP 中虚拟网关的地址可以和端口上的地址相同，VRRP 中端口只有 3 个状态：初始状态（Initial）、主状态（Master）和备份状态（Backup）。

VRRP 和 HSRP 比较如表 5-1 所示。

表 5-1　VRRP 和 HSRP 比较

HSRP	VRRP
是 Cisco 私有协议	是 IEEE 标准
最多支持 16 个组	最多支持 256 个组
一台活动路由器、一台备份路由器、其他是候选路由器	一台活动路由器、其他是备份路由器
虚拟 IP 地址不能和真实路由器的 IP 相同	虚拟 IP 地址能和真实路由器的 IP 相同
使用 224.0.0.2 地址发送消息	使用 224.0.0.18 地址发送消息
默认计时器：Hellotime 为 3 秒，Holdtime 为 10 秒	默认计时器：Advertisement Interval 为 1 秒，Master Down Interval 为 3 秒
可以 Track 端口	可以 Track 对象
支持明文和 MD5 认证	支持明文和 MD5 认证

VRRP 只有一种报文，VRRP 协议数据包格式如图 5-2 所示。

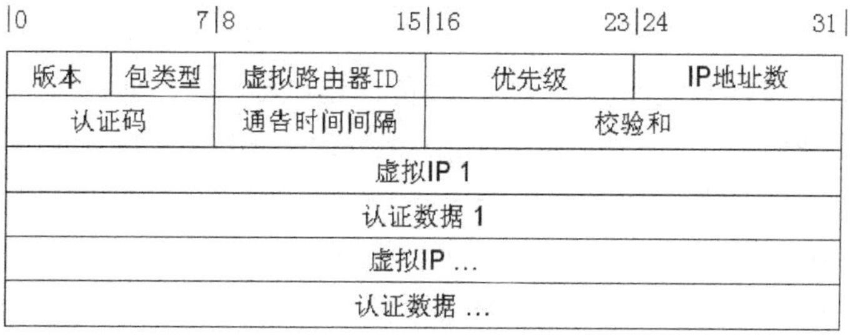

```
 0        7|8       15|16      23|24      31|
| 版本 | 包类型 | 虚拟路由器ID | 优先级 | IP地址数 |
|  认证码  |  通告时间间隔  |      校验和      |
|              虚拟IP 1              |
|            认证数据 1              |
|              虚拟IP ...            |
|            认证数据 ...            |
```

图 5-2　VRRP 协议数据包格式

① 版本：指示 VRRP 的版本信息。Cisco 默认支持 Version2。
② 包类型：只有一种类型，即 VRRP 通告报文（Advertisement），该字段取值为 1。
③ 虚拟路由器 ID：虚拟路由器号（即备份组号），取值范围为 1～255。
④ 优先级：路由器在备份组中的优先级，取值范围为 0～255，数值越大表明优先级越高。
⑤ IP 地址数：备份组虚拟 IP 地址的个数。1 个备份组可对应多个虚拟 IP 地址。

⑥ 认证码：该值为 0 表示无认证，为 1 表示简单字符认证，为 2 表示 MD5 认证。

⑦ 通告时间间隔：发送通告报文的时间间隔。在 VRRP Version2 中单位为秒，默认值为 1 秒。

⑧ 校验和：16 位校验和，用于检测 VRRP 报文中的数据破坏情况。

⑨ 虚拟 IP 地址：备份组虚拟 IP 地址表项。

⑩ 认证数据：验证字，目前只用于简单字符认证，对于其他认证方式一律填 0。

和 HSRP 一样，VRRP 根据优先级来确定备份组中每台路由器的角色（Master 路由器或 Backup 路由器）。优先级越高，则越有可能成为 Master 路由器。VRRP 优先级的取值范围为 0~255（数值越大表明优先级越高），可配置的范围是 1~254，优先级 0 为系统保留给特殊用途来使用，优先级 255 则是系统保留给 IP 地址拥有者使用。当路由器为 IP 地址拥有者时，其优先级始终为 255。因此，当备份组内存在 IP 地址拥有者时，只要其工作正常，则为 Master 路由器。

VRRP 定时器有如下 3 个。

① 通告时间间隔定时器（Advertisement Interval）：VRRP 备份组中的 Master 路由器会定时发送 VRRP 通告报文，通知备份组内的路由器自己工作正常。用户可以通过设置 VRRP 定时器来调整 Master 路由器发送 VRRP 通告报文的时间间隔。默认值为 1 秒。

② 时滞时间定时器：该值的计算方式为（256–优先级 / 256），单位为秒。

③ 主用失效时间间隔定时器（Master Down Interval）：如果 Backup 路由器在等待了 3 个间隔时间后，依然没有收到 VRRP 通告报文，则认为自己是 Master 路由器，并对外发送 VRRP 通告报文，重新进行 Master 路由器的选举。Backup 路由器并不会立即抢占成为 Master，而是等待一定时间（时滞时间）后，才会对外发送 VRRP 通告报文取代原来的 Master 路由器。因此该定时器值=3×通告时间间隔+（256–优先级 / 256）秒。

5.1.3 GLBP

HSRP 和 VRRP 能实现网关的冗余，然而如果要实现负载平衡，需要创建多个组，并让客户端指向不同的网关，这会给用户带来很大的不便。GLBP（Gateway Load Balance Protocol）也是 Cisco 的专有协议，不仅提供冗余网关功能，还在各网关之间提供负载均衡。GLBP 也是由多个路由器组成的一个组，并虚拟一个网关出来。GLBP 选举出一个 AVG（Active Virtual Gateway），AVG 不是负责转发数据的。AVG 分配 4 个 MAC 地址（每个 AVF 一个）给一个虚拟网关（Active Virtual Forwarder，AVF），并在计算机进行 ARP 请求时，根据负载平衡策略用不同的 MAC 地址进行响应，这样计算机实际就把数据发送给不同的 AVF 了，从而实现负载平衡。在 GLBP 中，真正负责转发数据的是 AVF，而不是 AVG，不过一个路由器可以同时是 AVG 和 AVF。

AVG 的选举和 HRSP 中活动路由器的选举非常类似，优先级最高的路由器为 AVG，次之的为 Backup AVG，其余的处于监听状态。一个 GLBP 组只能有一个 AVG 和一个 Backup AVG，主 AVG 失败，备份 AVG 顶上。

路由器是某些 MAC 地址（AVF）的活动路由器，GLBP 会控制 GLBP 组中哪个路由器是哪个 MAC 地址的活动路由器。如果计算机把数据发往某个 MAC 地址，该 MAC 地址的活动路由器将接收数据。当某一 MAC 的活动路由器发生故障时，其他路由器将成为这一 MAC 地址的新的活动路由器，从而实现冗余功能。

GLBP 的负载平衡策略可以根据不同主机简单地采取轮询方式或者根据路由器的权重平衡，默认采取轮询方式。

例 1：GLBP 组中有 3 台路由器 R1、R2、R3，R1 是 AVG，组中有 4 个 AVF，MAC 地址分别为 0007.b400.0101、0007.b400.0102、0007.b400.0103 和 0007.b400.0104。R1 是 0007.b400.0101 这个 AVF 的活动路由器；R2 是 0007.b400.0102 这个 AVF 的活动路由器；R3 是 0007.b400.0103 和 0007.b400.0104 这两个 AVF 的活动路由器，则如果采用轮询负载均衡，则 R1、R2 和 R3 实际分别承担 25%、25% 和 50% 的流量。

例 2：GLBP 组中有 6 台路由器 R1、R2、R3、R4、R5 和 R6，R1 是 AVG，组中还是只有 4 个 AVF，MAC 地址分别为 0007.b400.0101、0007.b400.0102、0007.b400.0103 和 0007.b400.0104。R1 是 0007.b400.0101 这个 AVF 的活动路由器；R2 是 0007.b400.0102 这个 AVF 的活动路由器；R3 是 0007.b400.0103 这个 AVF 的活动路由器；R4 是 0007.b400.0104 这个 AVF 的活动路由器；R5 和 R6 则处于监听状态。如果采用轮询负载均衡，则 R1、R2、R3 和 R4 实际各承担 25% 的流量。当 R4 发生故障时，R5 成为 0007.b400.0104 这个 AVF 的活动路由器，它将接替 R4，R6 仍处于监听状态。

5.1.4　SLB

SLB 全称为 Server Load Balancing，用于实现多个服务器之间的负载均衡。SLB 虚拟出一台服务器，对用户呈现的就是这个虚拟的服务器。虚拟服务器代表的是多个真实服务器的群集，当客户端向虚拟服务器发起连接时，SLB 通过某种均衡算法，转发到某真实服务器。负载均衡算法有两种：Weighted Round Robin（WRR）和 Weighted Least Connections（WLC），WRR 使用加权轮询算法分配连接，WLC 通过一定的权值，将下一个连接分配给活动连接数少的服务器。

SLB 有两种模式，第一种是分派模式（Dispatch），如图 5-3 所示。运行 SLB 的路由器收到用户计算机发来的数据包（其目的 IP 是虚拟服务器的 IP 地址）后，路由器会把数据包按照负载均衡算法分派到不同的真实服务器上，不会改变数据包中的目的 IP 地址。真实服务器收到的数据包中的目的 IP 地址是虚拟服务器的 IP 地址，因此需要在真实服务器上添加环回口并把环回口的地址设置为虚拟服务器的 IP 地址，或者添加第二个 IP 地址为虚拟服务器的 IP 地址，否则真实服务器会丢弃数据包。

SLB 的另一种模式是定向模式（Direct），如图 5-4 所示。运行 SLB 的路由器收到用户计算机发来的数据包（其目的 IP 是虚拟服务器的 IP 地址）后，路由器会把数据包也按照负载均衡算法分派到不同的真实服务器上，但是会把数据包中的目的 IP 地址改为真实服务器的 IP 地址。真实服务器收到的数据包中的目的 IP 地址是自己的 IP 地址，不会丢弃数据包。也就是说真实服务器并不需要知道虚拟服务器的存在。

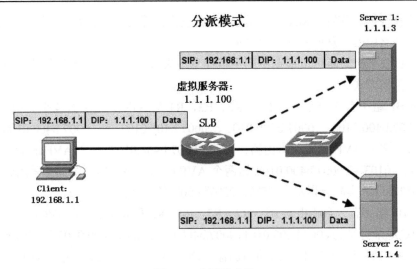

图 5-3　分派模式的 SLB

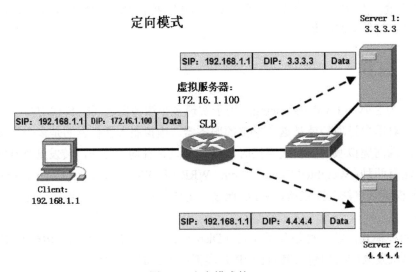

图 5-4　定向模式的 SLB

5.1.5　Syslog

Syslog 是网络设备向日志服务器发送日志的协议。网络设备根据网络事件所导致的结果生成日志消息，每个 Syslog 消息中包含一个严重级别，严重级别可以用数字表示（0～7），其含义如表 5-2 Syslog 安全级别所示。

表 5-2　Syslog 安全级别

严 重 级 别	含　　义	说　　明
0	emergencies	System is unusable（最高级别）
1	alerts	Immediate action needed

续表

严重级别	含义	说明
2	critical	Critical conditions
3	errors	Error conditions
4	warnings	Warning conditions
5	notifications	Normal but significant conditions
6	informational	Informational messages
7	debugging	Debugging messages

Syslog 消息格式如图 5-5 所示。

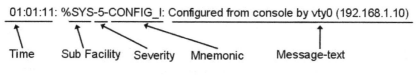

图 5-5　Syslog 消息格式

- Time：发生的时间；
- Sub Facility：表示硬件设备、协议或者系统软件的型号；
- Serverity：严重程度；
- Mnemonic：标识出错消息的代码；
- Message-text：消息体，可能会包含事件的具体信息，例如，端口号和网络地址。

5.1.6　SNMP

SNMP 提供了一种从网络上的设备中收集网络管理信息的方法。SNMP 也为设备向网络管理工作站报告问题和错误提供了一种方法。

使用 SNMP 进行网络管理需要下面几个重要组件：管理工作站、SNMP 代理、管理信息库（MIB）和网络管理工具。管理工作站通常就是计算机，管理工作站上必须装备有网络管理软件，管理员可以使用用户端口从 MIB 取得信息，同时为了进行网络管理，管理工作站应该具备将管理命令发到 SNMP 代理的能力。

SNMP 代理是一种网络设备，如主机、网桥和路由器等，这些设备都必须能够接收管理工作站发来的信息，它们的状态也必须可以由管理工作站监视。SNMP 代理响应管理工作站的请求进行相应的操作，也可以在没有请求的情况下向管理工作站发送信息。

MIB 是对象的集合，它代表网络中可以管理的资源和设备。每个对象基本上是一个数据变量，它代表被管理的对象的某一方面的信息。

最后一个组件是管理协议（网络管理工具），也就是 SNMP，SNMP 的基本功能是取得、设置和接收 SNMP 代理发送的意外信息。取得指的是管理工作站发送请求，SNMP 代理根据这个请求回送相应的数据；设置是管理工作站设置管理对象（也就是代理）的值；接收 SNMP 代理发送的意外信息是指代理可以在管理工作站未请求的状态下向管理工作站报告发生的意外情况。

5.1.7 交换机堆叠

思科的一些型号交换机支持堆叠，例如，Catalyst 3750 系列使用 StackWise 技术，提供了一个 32 Gbps 的堆叠互联，连接多达 9 台交换机。堆叠将交换机整合为一个统一的、逻辑的设备，交换机堆叠后从逻辑上来说，它们属于同一个设备。如果想对这几台交换机进行设置，只要连接到任何一台设备上，就可看到堆叠中的其他交换机。Catalyst 3750 采用的是背板堆叠方式，交换机背后有专门的堆叠口并且需专门的堆叠线。

对交换机进行堆叠时最好能够使用相同型号的硬件设备和相同版本的 IOS。如果硬件型号不同，则务必保证交换机上的 IOS 中使用的堆叠协议的主版本号是相同的；如果使用的堆叠协议的主版本号相同，但小数点后的版本号不同，仍然可以堆叠，不过可能会导致部分功能不兼容。通常堆叠的连接方式如图 5-6 所示，也可以如图 5-7 所示，这两种堆叠都能提供全带宽的功能，并且提供冗余功能。如图 5-7 所示，如果 B 链路出现故障，交换机堆叠仍能工作，但此时提供半带宽功能。而采用如图 5-8 所示的连接方式，如果 B 链路出现故障，交换机堆叠就被拆成两个堆叠了，这两个堆叠不能连通。

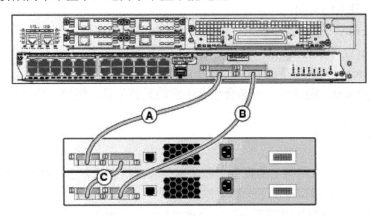

图 5-6　交换机堆叠连接方式 1

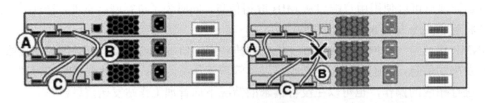

图 5-7　交换机堆叠连接方式 2

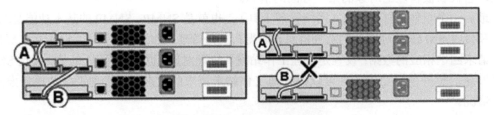

图 5-8　交换机堆叠连接方式 3

堆叠的交换机会选举一个 Master 设备，高优先级的设备是 Master，如果优先级相同，则会根据复杂的因素选举 Master（例如，IOS 的功能、版本、硬件版本和 MAC 地址等）。其他交换机则是成员（Member），如果 Master 交换机出现故障，则重新选举 Master。

5.2 实验 1：HSRP

1. 实验目的

通过本实验，读者可以掌握：
① HSRP 的工作原理。
② HSRP 的配置。

2. 实验 1 和实验 2 实验拓扑

实验 1 和实验 2 实验拓扑如图 5-9 所示。

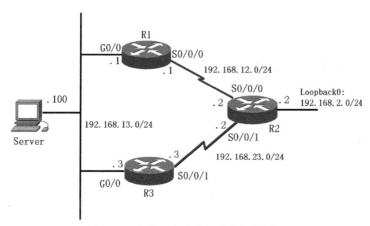

图 5-9　实验 1 和实验 2 实验拓扑图

在路由器和三层交换机上都可以配置 HSRP，这里是在路由器上配置。

3. 实验步骤

（1）配置 IP 地址和路由协议等

```
R1(config)#interface GigabitEthernet0/0
R1(config-if)#no shutdown
R1(config-if)#ip address 192.168.13.1 255.255.255.0
R1(config)#interface Serial0/0/0
R1(config-if)#no shutdown
R1(config-if)#ip address 192.168.12.1 255.255.255.0
R1(config)#router rip
R1(config-router)#network 192.168.12.0
R1(config-router)#network 192.168.13.0
```

R1(config-router)#**passive-interface GigabitEthernet0/0**
//之所以把 G0/0 端口设为被动端口，是防止从该端口发送 RIP 信息给 R3

R2(config)#**interface loopback 0**
R2(config-if)#**no shutdown**
R2(config-if)#**ip address 192.168.2.2 255.255.255.0**
R2(config)#**interface Serial0/0/0**
R2(config-if)#**no shutdown**
R2(config-if)#**ip address 192.168.12.2 255.255.255.0**
R2(config)#**interface Serial0/0/1**
R2(config-if)#**no shutdown**
R2(config-if)#**ip address 192.168.23.2 255.255.255.0**
R2(config)#**router rip**
R2(config-router)#**network 192.168.12.0**
R2(config-router)#**network 192.168.23.0**
R2(config-router)#**network 192.168.2.0**

R3(config)#**interface GigabitEthernet0/0**
R3(config-if)#**no shutdown**
R3(config-if)#**ip address 192.168.13.3 255.255.255.0**
R3(config)#**interface Serial0/0/1**
R3(config-if)#**no shutdown**
R3(config-if)#**ip address 192.168.23.3 255.255.255.0**
R3(config)#**router rip**
R3(config-router)#**network 192.168.23.0**
R3(config-router)#**network 192.168.13.0**
R3(config-router)#**passive-interface GigabitEthernet0/0**

（2）配置 HSRP

R1(config)#**interface GigabitEthernet0/0**
R1(config-if)#**standby 1 ip 192.168.13.254**
//启用 HSRP 功能并设置虚拟 IP 地址，1 为 standby 的组号。相同组号的路由器属于同一个 HSRP 组，所有属于同一个 HSRP 组的路由器的虚拟地址必须一致

R1(config-if)#**standby 1 priority 120**
//配置 HSRP 的优先级，如果不设置该项，默认优先级为 100，该值大的路由器会抢占成为活动路由器
R1(config-if)#**standby 1 preempt**
//该设置允许该路由器在优先级是最高时成为活动路由器。如果不设置，即使该路由器权值再高，也不会成为活动路由器。如果使用 "**standby 1 preempt delay minimum 1000**" 命令，则会延时 1 000 毫秒才进行抢占

R1(config-if)#**standby 1 timers 3 10**
//其中 3（秒）为 Hellotime，表示路由器每间隔多长时间发送 Hello 信息。10（秒）为 Holdtime，表示在多长时间内同组的其他路由器没有收到活动路由器的信息，则认为活动路由器发生故障。该设置的默认值分别为 3 秒和 10 秒。如果要更改默认值，所有同 HSRP 组的路由器的该项设置必须一致

R1(config-if)#**standby 1 authentication md5 key-string cisco**

第 5 章 高可用性

//配置认证密码，防止非法设备加入到 HSRP 组中，同一个组的密码必须一致

R3(config)#**interface GigabitEthernet0/0**
R3(config-if)#**standby 1 ip 192.168.13.254**
R3(config-if)#**standby 1 preempt**
R3(config-if)#**standby 1 timers 3 10**
R3(config-if)#**standby 1 authentication md5 key-string cisco**
//R2 上没有配置优先级，默认为 100

（3）检查、测试 HSRP

R1#**show standby brief**
```
                     P indicates configured to preempt.
                     |
Interface   Grp   Pri P State     Active          Standby         Virtual IP
Gi0/0       1     120 P Active    local           192.168.13.3    192.168.13.254
```
//以上表明 R1 的 G0/0 端口在 HSRP 组 1 中，当前是活动路由器，备份路由器为 192.168.13.3，虚拟 IP 地址是 192.168.13.254。

R1#**show standby**
//以下命令显示详细的信息
```
GigabitEthernet0/0 - Group 1
  State is Active
    2 state changes, last state change 00:00:33      //状态变化了 2 次
    Virtual IP address is 192.168.13.254              //虚拟 IP 地址
    Active virtual MAC address is 0000.0c07.ac01
```
//虚拟网关的 MAC 地址，默认是 0000.0c07.acXX，XX 是 HSRP 的组号。可以使用命令"**standby 1 mac-address**"修改虚拟网关的 MAC 地址
```
    Local virtual MAC address is 0000.0c07.ac01 (v1 default)
    Hello time 3 sec, hold time 10 sec                //计时器值
    Next hello sent in 1.632 secs
    Authentication MD5, key-string                    //认证方式
    Preemption enabled                                //启用抢占
    Active router is local                            //本地路由器就是活动路由器
    Standby router is 192.168.13.3, priority 100 (expires in 10.784 sec)
    Priority 120 (configured 120)                    //优先级为 120
    Group name is "hsrp-Gi0/0-1" (default)
```
//HSRP 组名，可以使用命令"**standby 1 name**"设置，如果没有设置，则系统自动生成

R3#**show standby brief**
```
                     P indicates configured to preempt.
                     |
Interface   Grp   Pri P State     Active          Standby         Virtual IP
G0/0        1     100 P Standby   192.168.13.1    local           192.168.13.254
```
//表明 R3 是备份路由器，活动路由器为 192.168.13.1

在 Server 上配置的 IP 地址为 192.168.13.100/24，网关指向 192.168.13.254。在 Server 上连续 ping 路由器 R2 的 Loopback0 端口（192.168.2.2），在 R1 上关闭 G0/0 端口，观察 Server 上 ping 的结果。如下所示：

```
C:\>ping -t 192.168.2.2
Reply from 192.168.2.2: bytes=32 time=9ms TTL=254
Reply from 192.168.2.2: bytes=32 time=9ms TTL=254
Reply from 192.168.2.2: bytes=32 time=9ms TTL=254
Request timed out.
Reply from 192.168.2.2: bytes=32 time=9ms TTL=254
Reply from 192.168.2.2: bytes=32 time=9ms TTL=254
Reply from 192.168.2.2: bytes=32 time=11ms TTL=254
Reply from 192.168.2.2: bytes=32 time=9ms TTL=254
//当 R1 故障时，R3 很快就替代了 R1，计算机的通信只受到短暂的影响

R3#show standby  brief
                     P indicates configured to preempt.
                      |
Interface Grp   Pri P State    Active           Standby         Virtual IP
Gi0/0       1   100 P Active   local            unknown         192.168.13.254
//表明 R3 成为活动路由器了
```

提示

如果重新打开 R1 的 G0/0 端口，由于在 R1 上配置抢占（Preempt），因此 R1 又会成为活动路由器。

（4）Debug 信息

以下是当路由器 R1 的 G0/0 端口关闭后，R3 成为活动路由器过程中的 debug 信息：

```
R3#debug standby
*Feb   4 02:08:50.927: HSRP: Gi0/0 Interface adv out, Passive, active 0 passive 1
*Feb   4 02:08:52.891: HSRP: Gi0/0 Grp 1 Hello   in  192.168.13.1 Active   pri 120 vIP 192.168.13.254
//收到 R1 发送的 Hello 包
*Feb   4 02:08:53.323: HSRP: Gi0/0 Grp 1 Hello   out 192.168.13.3 Standby pri 100 vIP 192.168.13.254
// R3 发送 Hello 包
*Feb   4 02:08:55.091: HSRP: Gi0/0 Grp 1 Resign in  192.168.13.1 Active   pri 120 vIP 192.168.13.254
*Feb   4 02:08:55.091: HSRP: Gi0/0 Grp 1 Standby: i/Resign rcvd (120/192.168.13.1)
*Feb   4 02:08:55.091: HSRP: Gi0/0 Grp 1 Active router is local, was 192.168.13.1
*Feb   4 02:08:55.091: HSRP: Gi0/0 Nbr 192.168.13.1 no longer active for group 1 (Standby)
// R1（192.168.13.1）不再是 Active 路由器了
*Feb   4 02:08:55.091: HSRP: Gi0/0 Nbr 192.168.13.1 Was active or standby - start passive holddown
// R3 开始切换成为活动路由器
*Feb   4 02:08:55.091: HSRP: Gi0/0 Grp 1 Standby router is unknown, was local
//因为 R1 发生故障，所以没有备份路由器了
*Feb   4 02:08:55.091: HSRP: Gi0/0 Grp 1 Standby -> Active
*Feb   4 02:08:55.091: HSRP: Gi0/0 Interface adv out, Active, active 1 passive 0
```

```
*Feb  4 02:08:55.091: %HSRP-5-STATECHANGE: GigabitEthernet0/0 Grp 1 state Standby -> Active
*Feb  4 02:08:55.091: HSRP: Gi0/0 Grp 1 Redundancy "hsrp-Gi0/0-1" state Standby -> Active
*Feb  4 02:08:55.091: HSRP: Gi0/0 Grp 1 Hello    out 192.168.13.3 Active    pri 100 vIP 192.168.13.254
*Feb  4 02:08:55.091: HSRP: Gi0/0 Grp 1 Added 192.168.13.254 to ARP (0000.0c07.ac01)
*Feb  4 02:08:55.091: HSRP: Gi0/0 Grp 1 Activating MAC 0000.0c07.ac01
*Feb  4 02:08:55.091: HSRP: Gi0/0 Grp 1 Adding 0000.0c07.ac01 to MAC address filter
*Feb  4 02:08:55.091: HSRP: Gi0/0 IP Redundancy "hsrp-Gi0/0-1" standby, local -> unknown
*Feb  4 02:08:55.091: HSRP: Gi0/0 IP Redundancy "hsrp-Gi0/0-1" update, Standby -> Active
*Feb  4 02:08:57.979: HSRP: Gi0/0 Grp 1 Hello    out 192.168.13.3 Active    pri 100 vIP 192.168.13.254
*Feb  4 02:08:58.091: HSRP: Gi0/0 IP Redundancy "hsrp-Gi0/0-1" update, Active -> Active
//以上是路由器的 HSRP debug 信息
*Feb  4 02:09:00.811: HSRP: Gi0/0 Grp 1 Hello    out 192.168.13.3 Active    pri 100 vIP 192.168.13.254
```

（5）配置端口跟踪

在图 5-9 中，按照以上步骤的配置，如果 R1 的 S0/0/0 端口出现问题，R1 将没有到达 R2 的 Loopback0 端口所在网段的路由。然而 R1 和 R3 之间的以太网仍然没有问题，HSRP 的 Hello 包可正常发送和接收，因此 R1 仍然是虚拟网关 192.168.13.254 的活动路由器，Server 的数据会发送给 R1，这样会造成 Server 无法 ping 通 R2 的 Loopback0 端口。我们可以配置端口跟踪解决这个问题，端口跟踪使得 R1 发现 S0/0/0 上的链路出现问题后，把自己的优先级（我们设为了 120）减去一个数字（如 30），成为了 90。由于 R3 的优先级为默认值为 100，R3 就成为了活动路由器。配置如下：

```
R1(config)#track 100 interface Serial0/0/0 line-protocol
R1(config)#int GigabitEthernet 0/0
R1(config-if)#standby 1 track 100 decrement 30
```

以上首先使用 **track** 命令定义了一个跟踪目标 100，该目标就是 S0/0/0。如果该端口发生故障，优先级降低 30，变为 120-30=90，这时，R3 的优先级为 100（默认值），因此 R3 会成为活动路由器，Server 数据包将发往 R3。降低的值应该选取合适的值，使得其他路由器能成为活动路由器。按照步骤（3）测试 HSRP 的端口跟踪是否生效。

（6）配置多个 HSRP 组

之前的步骤已经虚拟了 192.168.13.254 网关，对于这个网关只能有一个活动路由器，于是这个路由器将承担全部的数据流量。我们可以再创建一个 HSRP 组，虚拟出另一个网关 192.168.13.253，这时 R3 是活动路由器，让一部分计算机指向这个网关。这样就能做到负载平衡。以下是有 2 个 HSRP 组的完整配置。

在 R1 上：

```
track 100 interface Serial0/0/0 line-protocol
interface GigabitEthernet0/0
    standby 1 ip 192.168.13.254
    standby 1 priority 120
    standby 1 preempt
    standby 1 authentication md5 key-string cisco
    standby 2 ip 192.168.13.253
```

```
        standby 2 preempt
        standby 2 authentication md5 key-string cisco
        standby 1 track 100 decrement 30
```
在 R3 上：
```
track 100 interface Serial0/0/1 line-protocol
interface GigabitEthernet0/0
    standby 1 ip 192.168.13.254
    standby 1 preempt
    standby 1 authentication md5 key-string cisco
    standby 2 ip 192.168.13.253
    standby 2 priority 120
    standby 2 preempt
    standby 2 authentication md5 key-string cisco
    standby 2 track 100 decrement 30
```

技术要点

我们这里创建了两个 HSRP 组，第一个组的 IP 地址为 192.168.13.254，活动路由器为 R1，一部分计算机的网关指向 192.168.13.254。第二个组的 IP 地址为 192.168.13.253，活动路由器为 R2，另一部分计算机的网关指向 192.168.13.253。这样，当网络全部正常时，一部分数据是由 R1 转发的，另一部分数据是由 R2 转发的，实现了负载平衡。当一个路由器出现问题时，则另一个路由器就成为两个 HSRP 组的活动路由器，承担全部的数据转发功能。通过这种方式实现负载平衡，要求不同的计算机在设置网关时有所不同，如果计算机的 IP 地址是 DHCP 分配的，就不太方便。

技术要点

实际上 HSRP 在局域网用得较多，由于局域网内大多使用三层交换机，所以这时 HSRP 是在三层交换机上的 VLAN 端口配置的。另外，HSRP 有 Version1 和 Version2 两个版本，两者不兼容，因此所有路由器或者交换机要使用相同的 HSRP 版本，使用 **"standby version"** 命令设置。

5.3 实验 2：VRRP

1. 实验目的

通过本实验，读者可以掌握：
① VRRP 的工作原理。
② VRRP 的配置。

2. 实验拓扑

实验拓扑如图 5-9 所示。

3. 实验步骤

VRRP 的配置和 HSRP 的配置非常相同，有些配置不再重复说明了。

（1）配置 IP 地址和路由协议等

```
R1(config)#interface GigabitEthernet0/0
R1(config-if)#no shutdown
R1(config-if)#ip address 192.168.13.1 255.255.255.0
R1(config)#interface Serial0/0/0
R1(config-if)#no shutdown
R1(config-if)#ip address 192.168.12.1 255.255.255.0
R1(config)#router rip
R1(config-router)#network 192.168.12.0
R1(config-router)#network 192.168.13.0
R1(config-router)#passive-interface GigabitEthernet0/0
//之所以把 G0/0 端口设为被动端口，是防止从该端口发送 RIP 信息给 R3


R2(config)#interface loopback 0
R2(config-if)#no shutdown
R2(config-if)#ip address 192.168.2.2 255.255.255.0
R2(config)#interface Serial0/0/0
R2(config-if)#no shutdown
R2(config-if)#clock rate 128000
R2(config-if)#ip address 192.168.12.2 255.255.255.0
R2(config)#interface Serial0/0/1
R2(config-if)#no shutdown
R2(config-if)#clock rate 128000
R2(config-if)#ip address 192.168.23.2 255.255.255.0
R2(config)#router rip
R2(config-router)#network 192.168.12.0
R2(config-router)#network 192.168.23.0
R2(config-router)#network 192.168.2.0


R3(config)#interface GigabitEthernet0/0
R3(config-if)#no shutdown
R3(config-if)#ip address 192.168.13.3 255.255.255.0
R3(config)#interface Serial0/0/1
R3(config-if)#no shutdown
R3(config-if)#ip address 192.168.23.3 255.255.255.0
R3(config)#router rip
R3(config-router)#network 192.168.23.0
R3(config-router)#network 192.168.13.0
R3(config-router)#passive-interface GigabitEthernet0/0
```

（2）配置多个 VRRP 组并跟踪端口

在 R1 上：

```
R1(config)#track 100 interface Serial0/0/0 line-protocol
```

R1(config)#**interface GigabitEthernet0/0**
R1(config-if)#**vrrp 1 ip 192.168.13.254**
R1(config-if)#**vrrp 1 priority 120**
R1(config-if)#**vrrp 1 preempt**
R1(config-if)#**vrrp 1 authentication md5 key-string cisco**
R1(config-if)#**vrrp 1 track 100 decrement 30**
R1(config-if)#**vrrp 2 ip 192.168.13.253**
R1(config-if)#**vrrp 2 preempt**
R1(config-if)#**vrrp 2 authentication md5 key-string cisco**
//VRRP 端口跟踪和 HSRP 有些不同，需要在全局配置模式下先定义跟踪目标，再配置 VRRP 中跟踪该目标，我们这里定义了目标 100 是跟踪 S0/0/0 端口

在 R3 上：
R3config)#**track 100 interface Serial0/0/1 line-protocol**
R3(config)#**interface GigabitEthernet0/0**
R3(config-if)#**vrrp 1 ip 192.168.13.254**
R3(config-if)#**vrrp 1 preempt**
R3(config-if)#**vrrp 1 authentication md5 key-string cisco**
R3(config-if)#**vrrp 2 ip 192.168.13.253**
R3(config-if)#**vrrp 2 priority 120**
R3(config-if)#**vrrp 2 preempt**
R3(config-if)#**vrrp 2 authentication md5 key-string cisco**
R3(config-if)#**vrrp 2 track 100 decrement 30**

R1#**show vrrp brief**

Interface	Grp	Pri	Time	Own	Pre	State	Master addr	Group addr
Gi0/0	1	120	3531		Y	Master	192.168.13.1	192.168.13.254
Gi0/0	2	100	3609		Y	Backup	192.168.13.3	192.168.13.253

//以上表明 R1 是 192.168.13.254 虚拟网关的 Master 路由器，是 192.168.13.253 虚拟网关的 Backup 路由器

R1#**show vrrp**
GigabitEthernet0/0 - Group 1   //以下是组 1 的情况
  State is Master   //当前路由器是 Master
  Virtual IP address is 192.168.13.254   //虚拟 IP 地址
  Virtual MAC address is 0000.5e00.0101   //虚拟 IP 地址的 MAC 地址
  Advertisement interval is 1.000 sec   //本地路由器配置的通告时间间隔，默认值为 1 秒
  Preemption enabled   //已经开启抢占
  Priority is 120   //优先级为 120
    Track object 100 state Up decrement 30 //该组跟踪对象 100，当对象发生故障时优先级减 30
  Authentication MD5, key-string   //采用 MD5 认证
  Master Router is 192.168.13.1 (local), priority is 120   //显示 Master 路由器的信息
  Master Advertisement interval is 1.000 sec   //Master 路由器上的通告时间间隔
  Master Down interval is 3.531 sec
// Master 路由器上主用失效时间，等于 3×1+(256-120)/256=3.531 秒
GigabitEthernet0/0 - Group 2

```
State is Backup
Virtual IP address is 192.168.13.253
Virtual MAC address is 0000.5e00.0102
Advertisement interval is 1.000 sec
Preemption enabled
Priority is 100
Authentication MD5, key-string
Master Router is 192.168.13.3, priority is 120
Master Advertisement interval is 1.000 sec
Master Down interval is 3.609 sec (expires in 3.241 sec)

R3#show vrrp brief
Interface        Grp  Pri Time   Own Pre  State    Master addr      Group addr
Gi0/0            1    100 3609       Y    Backup   192.168.13.1     192.168.13.254
Gi0/0            2    120 3531       Y    Master   192.168.13.3     192.168.13.253
```
//以上表明 R3 是 192.168.13.253 虚拟网关的 Master 路由器，是 192.168.13.254 虚拟网关的 Backup 路由器

（3）测试

在 Server 上配置 IP 地址 192.168.13.100/24，网关指向 192.168.13.254。在 Server 上连续 ping 路由器 R2 的 Loopback0 端口（192.168.2.2），在 R1 上关闭 G0/0 端口，观察 Server 上 ping 的结果。如下所示：

```
C:\>ping -t 192.168.2.2
Reply from 192.168.2.2: bytes=32 time=9ms TTL=254
Reply from 192.168.2.2: bytes=32 time=9ms TTL=254
Reply from 192.168.2.2: bytes=32 time=9ms TTL=254
Request timed out.
Reply from 192.168.2.2: bytes=32 time=9ms TTL=254
Reply from 192.168.2.2: bytes=32 time=9ms TTL=254
Reply from 192.168.2.2: bytes=32 time=11ms TTL=254
Reply from 192.168.2.2: bytes=32 time=9ms TTL=254
```
//从以上可以看到，当 R1 发生故障时，R3 很快就替代了 R1，计算机的通信只受到短暂的影响

（4）Debug 信息

以下是路由器 R1 的 G0/0 端口关闭后，R3 成为活动路由器过程中的 debug 信息。

```
R3#debug vrrp
*Feb  4 02:41:37.355: VRRP: Sent: 21027801FE010000C0A80DFD0000000000000000
*Feb  4 02:41:37.355: VRRP: HshC: E908B5C08E257F0ACD02FEC1F15A65FF
*Feb  4 02:41:37.355: VRRP: Grp 2 sending Advertisement checksum FE72
*Feb  4 02:41:37.975:
*Feb  4 02:41:37.975: VRRP: Rcvd: 21010001FE010000C0A80DFE0000000000000000
*Feb  4 02:41:37.975: VRRP: HshC: 5DE47364BAAD6A2C2BA2C4135651EDC4
*Feb  4 02:41:37.975: VRRP: HshR: 5DE47364BAAD6A2C2BA2C4135651EDC4
*Feb  4 02:41:37.975: VRRP: Grp 1 Advertisement priority 0, ipaddr 192.168.13.1
```

```
*Feb  4 02:41:37.975: VRRP: Grp 1 Event - Advert priority 0
*Feb  4 02:41:38.211:
*Feb  4 02:41:38.211: VRRP: Sent: 21027801FE010000C0A80DFD0000000000000000
*Feb  4 02:41:38.211: VRRP: HshC: E908B5C08E257F0ACD02FEC1F15A65FF
*Feb  4 02:41:38.211: VRRP: Grp 2 sending Advertisement checksum FE72
*Feb  4 02:41:38.587: VRRP: Grp 1 Event - Master down timer expired
*Feb  4 02:41:38.587: %VRRP-6-STATECHANGE: Gi0/0 Grp 1 state Backup -> Master
*Feb  4 02:41:38.587: VRRP: tbridge_smf_update failed
//R3 成为 VRRP 组 1 的 Master
```

5.4 实验 3：GLBP

1. 实验目的

通过本实验，读者可以掌握：
① GLBP 的工作原理。
② GLBP 的配置。

2. 实验拓扑

GLBP 实验拓扑如图 5-10 所示。

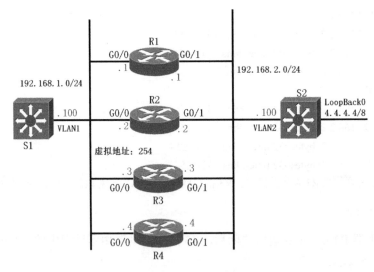

图 5-10　GLBP 实验拓扑

交换机 S1 作为计算机使用，用于测试。在交换机 S2 上启用路由功能，并且创建环回口，该环回口用于测试。

3. 实验步骤

（1）划分 VLAN，配置 IP 地址和路由协议等

```
R1(config)#interface GigabitEthernet0/0
```

```
R1(config-if)#no shutdown
R1(config-if)#ip address 192.168.1.1 255.255.255.0
R1(config)#interface GigabitEthernet0/1
R1(config-if)#no shutdown
R1(config-if)#ip address 192.168.2.1 255.255.255.0
R1(config)#router rip
R1(config-router)#network 192.168.1.0
R1(config-router)#network 192.168.2.0
R1(config-router)#passive-interface GigabitEthernet0/0

R2(config)#interface GigabitEthernet0/0
R2(config-if)#no shutdown
R2(config-if)#ip address 192.168.1.2 255.255.255.0
R2(config)#interface GigabitEthernet0/1
R2(config-if)#no shutdown
R2(config-if)#ip address 192.168.2.2 255.255.255.0
R2(config)#router rip
R2(config-router)#network 192.168.1.0
R2(config-router)#network 192.168.2.0
R2(config-router)#passive-interface GigabitEthernet0/0

R3(config)#interface GigabitEthernet0/0
R3(config-if)#no shutdown
R3(config-if)#ip address 192.168.1.3 255.255.255.0
R3(config)#interface GigabitEthernet0/1
R3(config-if)#no shutdown
R3(config-if)#ip address 192.168.2.3 255.255.255.0
R3(config)#router rip
R3(config-router)#network 192.168.1.0
R3(config-router)#network 192.168.2.0
R3(config-router)#passive-interface GigabitEthernet0/0

R4(config)#interface GigabitEthernet0/0
R4(config-if)#no shutdown
R4(config-if)#ip address 192.168.1.4 255.255.255.0
R4(config)#interface GigabitEthernet0/1
R4(config-if)#no shutdown
R4(config-if)#ip address 192.168.2.4 255.255.255.0
R4(config)#router rip
R4(config-router)#network 192.168.1.0
R4(config-router)#network 192.168.2.0
R4(config-router)#passive-interface GigabitEthernet0/0

S1(config)#interface range FastEthernet 0/22 -24
```

```
S1(config-if-range)#shutdown
//关闭 S1 与 S2 和 S3 交换机之间的连接链路
S1(config-if-range)#interface vlan 1
S1(config-if)#no shutdown
S1(config-if)#ip address 192.168.1.100 255.255.255.0
//配置 VLAN1 端口的 IP 地址
S1(config)#ip default-gateway 192.168.1.254
//配置 VLAN1 的网关，该网关就是 GLBP 虚拟出的网关


S2(config)#interface range FastEthernet 0/22 -24
S2(config-if-range)#shutdown
S2(config)#interface range FastEthernet 0/1 -4
S2(config-if-range)#switchport mode access
S2(config-if-range)#switchport access vlan 2
//把端口划分到 VLAN 2
S2(config)#ip routing
//启用路由功能
S2(config)#interface vlan 2
S2(config-if)#no shutdown
S2(config-if)#ip address 192.168.2.100 255.255.255.0
S2(config-if)#int loopback0
S2(config-if)#ip address 4.4.4.4 255.0.0.0
S2(config-if)#exit
S2(config-if)#router rip
S2(config-router)#network 192.168.2.0
S2(config-router)#network 4.0.0.0
//配置路由协议
```

（2）配置 GLBP

```
R1(config)#track 100 interface GigabitEthernet0/1 line-protocol
//配置跟踪目标 100 是 G0/1 端口的二层故障
R1(config)#interface GigabitEthernet0/0
R1(config-if)#glbp 1 ip 192.168.1.254
//和 HSRP 类似，以上命令创建 GLBP 组，虚拟网关的 IP 为 192.168.1.254
R1(config-if)#glbp 1 name TEST
//配置一个名字，这是可选的
R1(config-if)#glbp 1 authentication md5 key-string cisco
//配置认证，防止非法设备接入
R1(config-if)#glbp 1 timers 3 10
//配置 GLDP 的 Hello 时间和 Hold 时间，默认值分别为 3 秒和 10 秒
R1(config-if)#glbp 1 priority 200
//配置优先级，优先级高的路由器成为 AVG，默认值为 100。注意：这个优先级不是控制 AVF 的选举，而是控制 AVG 的选举
R1(config-if)#glbp 1    preempt
```

//配置路由器会进行 AVG 抢占，否则即使优先级再高，也不会成为 AVG
R1(config-if)#**glbp 1 preempt delay minimum 1**
//配置 AVG 抢占延时为 1 秒
R1(config-if)#**glbp 1 weighting 200 lower 170 upper 180**

以上配置路由器的权重为 200，低限为 170，高限为 180。这个值很重要，当路由器出现故障后，如果权重低于低限（170）时，该路由器将不会成为 AVF，也就是不会转发数据；当故障排除后，如果权重高于高限（180），路由器重新成为 AVF，才能转发数据。此外，如果采用基于权重的负载均衡，则权重越大，转发数据的机会越多。

R1(config-if)#**glbp 1 weighting track 100　decrement 50**

以上命令配置跟踪目标 100，即跟踪 G0/1 端口，如果 G0/1 端口发生故障则权重减去 50。由于路由器的权重为 200，G0/1 端口如发生故障，则权重为 200-50=150，该值小于 170，路由器不转发数据。当 G0/1 端口恢复时，权重恢复为 200，高于 180，路由器恢复转发数据的能力。可以配置多个跟踪目标。

R1(config-if)#**glbp 1 forwarder preempt**
//配置该路由器会抢占成为 AVF
R1(config-if)#**glbp 1 forwarder preempt delay minimum 25**
//配置该路由器抢占成为 AVF 的延迟时间为 25 秒，默认值为 30 秒
R1(config-if)#**glbp 1 load-balancing round-robin**
//配置均衡策略，默认值就是 round-robin

技术要点

默认时 GLBP 的负载均衡策略是轮询方式，可以在端口下使用"**glbp 1 load-balancing**"命令修改，有以下选项。

① host-dependent：根据不同主机的源 MAC 地址进行平衡。
② round-robin：轮询方式，即每响应一次 ARP 请求，轮换一个地址。
③ weighted：根据路由器的权重分配，权重高的被分配的可能性越大。

R2(config)#**track 100 interface GigabitEthernet0/1 line-protocol**
R2(config)#**int GigabitEthernet0/0**
R2(config-if)#**glbp 1 ip 192.168.1.254**
R2(config-if)#**glbp 1 name TEST**
R2(config-if)#**glbp 1 authentication md5 key-string cisco**
R2(config-if)#**glbp 1 timers 3 10**
R2(config-if)#**glbp 1 priority 180**
//R2 的优先级为 180，比 R1 小
R2(config-if)#**glbp 1　preempt**
R2(config-if)#**glbp 1 preempt delay minimum 1**
R2(config-if)#**glbp 1 weighting 200 lower 170 upper 180**
R2(config-if)#**glbp 1 weighting track 100　decrement 50**
R2(config-if)#**glbp 1 forwarder preempt**
R2(config-if)#**glbp 1 forwarder preempt delay minimum 25**
R2(config-if)#**glbp 1 load-balancing round-robin**

R3(config)#**track 100 interface GigabitEthernet0/1 line-protocol**

```
R3(config)#int GigabitEthernet0/0
R3(config-if)#glbp 1 ip 192.168.1.254
R3(config-if)#glbp 1 name TEST
R3(config-if)#glbp 1 authentication md5 key-string cisco
R3(config-if)#glbp 1 timers 3 10
R3(config-if)#glbp 1 priority 160
//R3 的优先级为 160
R3(config-if)#glbp 1    preempt
R3(config-if)#glbp 1 preempt delay minimum 1
R3(config-if)#glbp 1 weighting 200 lower 170 upper 180
R3(config-if)#glbp 1 weighting track 100    decrement 50
R3(config-if)#glbp 1 forwarder preempt
R3(config-if)#glbp 1 forwarder preempt delay minimum 25
R3(config-if)#glbp 1 load-balancing round-robin


R4(config)#track 100 interface GigabitEthernet0/1 line-protocol
R4(config)#int GigabitEthernet0/0
R4(config-if)#glbp 1 ip 192.168.1.254
R4(config-if)#glbp 1 name TEST
R4(config-if)#glbp 1 authentication md5 key-string cisco
R4(config-if)#glbp 1 timers 3 10
R4(config-if)#glbp 1 priority 140
//R4 的优先级为 140
R4(config-if)#glbp 1    preempt
R4(config-if)#glbp 1 preempt delay minimum 1
R4(config-if)#glbp 1 weighting 200 lower 170 upper 180
R4(config-if)#glbp 1 weighting track 100    decrement 50
R4(config-if)#glbp 1 forwarder preempt
R4(config-if)#glbp 1 forwarder preempt delay minimum 25
R4(config-if)#glbp 1 load-balancing round-robin
```

（3）查看 GLBP 信息

```
R1#show glbp
GigabitEthernet0/0 - Group 1
    State is Active                                    //该路由器处于活动状态，即为 AVG
      1 state change, last state change 00:07:08
    Virtual IP address is 192.168.1.254                //虚拟的网关 IP 地址
    Hello time 3 sec, hold time 10 sec                 //计时器
      Next hello sent in 2.880 secs
    Redirect time 600 sec, forwarder timeout 14400 sec
    Authentication MD5, key-string                     //采用了 MD5 认证
    Preemption enabled, min delay 1 sec                //路由器会进行 AVG 抢占
    Active is local                                    //活动 AVG 路由器是本路由器
    Standby is 192.168.1.2, priority 180 (expires in 9.280 sec)    //Standby 路由器的信息
```

Priority 200 (configured)
Weighting 200 (configured 200), thresholds: lower 170, upper 180   //权重值、低限和高限
  Track object 100 state Up decrement 50   //跟踪目标；目标发生故障后权重减少 50
Load balancing: round-robin   //负载均衡方式为轮询
IP redundancy name is "TEST"   //GLBP 的名字
Group members: //该组中的成员，即 4 台路由器
  f872.ea69.18b8 (192.168.1.3) authenticated
  f872.ea69.1c78 (192.168.1.2) authenticated
  f872.eac8.4f98 (192.168.1.4) authenticated
  f872.ead6.f4c8 (192.168.1.1) local
There are 4 forwarders (1 active)
Forwarder 1   //以下是第一个 AVF 的情况
  State is Active   //本路由器是第一个 AVF 的活动路由器
    1 state change, last state change 00:06:57
  MAC address is 0007.b400.0101 (default)   //AVF 的 MAC 地址
  Owner ID is f872.ead6.f4c8   //谁是该 AVF 的活动路由器（拥有者）
  Redirection enabled
  Preemption enabled, min delay 25 sec   //AVF 的抢占延时
  Active is local, weighting 200   //本路由器是第一个 AVF 的活动路由器，权重为 200
Forwarder 2   //本路由器不是第二个 AVF 的活动路由器，状态为 Listen
  State is Listen
  MAC address is 0007.b400.0102 (learnt)
  Owner ID is f872.ea69.1c78
  Redirection enabled, 599.296 sec remaining (maximum 600 sec)
  Time to live: 14399.296 sec (maximum 14400 sec)
  Preemption enabled, min delay 25 sec
  Active is 192.168.1.2 (primary), weighting 200 (expires in 9.312 sec)
  //192.168.1.2（R2）是第二个 AVF 的活动路由器
Forwarder 3
  State is Listen
  MAC address is 0007.b400.0103 (learnt)
  Owner ID is f872.ea69.18b8
  Redirection enabled, 599.456 sec remaining (maximum 600 sec)
  Time to live: 14399.456 sec (maximum 14400 sec)
  Preemption enabled, min delay 25 sec
  Active is 192.168.1.3 (primary), weighting 200 (expires in 9.568 sec)
  //192.168.1.3（R3）是第三个 AVF 的活动路由器
Forwarder 4
  State is Listen
  MAC address is 0007.b400.0104 (learnt)
  Owner ID is f872.eac8.4f98
  Redirection enabled, 598.560 sec remaining (maximum 600 sec)
  Time to live: 14398.560 sec (maximum 14400 sec)
  Preemption enabled, min delay 25 sec

Active is 192.168.1.4 (primary), weighting 200 (expires in 9.472 sec)
　　　//192.168.1.4（R4）是第四个 AVF 的活动路由器

R2#show glbp
GigabitEthernet0/0 - Group 1
　State is Standby
　　1 state change, last state change 00:13:27
　Virtual IP address is 192.168.1.254
　Hello time 3 sec, hold time 10 sec
　　Next hello sent in 0.096 secs
　Redirect time 600 sec, forwarder timeout 14400 sec
　Authentication MD5, key-string
　Preemption enabled, min delay 1 sec
　Active is 192.168.1.1, priority 200 (expires in 9.024 sec)
　Standby is local
　Priority 180 (configured)
　Weighting 200 (configured 200), thresholds: lower 170, upper 180
　　Track object 100 state Up decrement 50
　Load balancing: round-robin
　IP redundancy name is "TEST"
　Group members:
　　f872.ea69.18b8 (192.168.1.3) authenticated
　　f872.ea69.1c78 (192.168.1.2) local
　　f872.eac8.4f98 (192.168.1.4) authenticated
　　f872.ead6.f4c8 (192.168.1.1) authenticated
　There are 4 forwarders (1 active)
　Forwarder 1
　　State is Listen
　　MAC address is 0007.b400.0101 (learnt)
　　Owner ID is f872.ead6.f4c8
　　Time to live: 14397.728 sec (maximum 14400 sec)
　　Preemption enabled, min delay 25 sec
　　Active is 192.168.1.1 (primary), weighting 200 (expires in 8.352 sec)
　Forwarder 2
　　State is Active
　　　1 state change, last state change 00:13:30
　　MAC address is 0007.b400.0102 (default)
　　Owner ID is f872.ea69.1c78
　　Preemption enabled, min delay 25 sec
　　Active is local, weighting 200
　Forwarder 3
　　State is Listen
　　MAC address is 0007.b400.0103 (learnt)
　　Owner ID is f872.ea69.18b8

```
    Time to live: 14398.336 sec (maximum 14400 sec)
    Preemption enabled, min delay 25 sec
    Active is 192.168.1.3 (primary), weighting 200 (expires in 8.768 sec)
  Forwarder 4
    State is Listen
    MAC address is 0007.b400.0104 (learnt)
    Owner ID is f872.eac8.4f98
    Time to live: 14398.848 sec (maximum 14400 sec)
    Preemption enabled, min delay 25 sec
    Active is 192.168.1.4 (primary), weighting 200 (expires in 10.272 sec)
```

通过查看，可以知道：

- R1——0007.b400.0101 的活动路由器；
- R2——0007.b400.0102 的活动路由器；
- R3——0007.b400.0103 的活动路由器；
- R4——0007.b400.0104 的活动路由器。

（4）检查 GLBP 的负载平衡功能

在 S1 上配置 IP 地址，网关指向 192.168.1.254，并进行如下操作：

```
S1#clear arp-cache
//清除 ARP 表
S1#ping 4.4.4.4
S1#show ip arp
Protocol  Address          Age (min)  Hardware Addr    Type   Interface
Internet  192.168.1.100       -       d0c7.89ab.11c0   ARPA   Vlan1
Internet  192.168.1.254       0       0007.b400.0101   ARPA   Vlan1
//以上表明 S1 的 ARP 请求获得网关（192.168.1.254）的 MAC 地址为 00-07-b4-00-01-01

S1#clear arp-cache
S1#ping 4.4.4.4
S1#show ip arp
Protocol  Address          Age (min)  Hardware Addr    Type   Interface
Internet  192.168.1.100       -       d0c7.89ab.11c0   ARPA   Vlan1
Internet  192.168.1.254       0       0007.b400.0102   ARPA   Vlan1
//以上表明 S1 的 ARP 请求获得网关（192.168.1.254）的 MAC 地址为 00-07-b4-00-01-02

S1#clear arp-cache
S1#ping 4.4.4.4
S1#show ip arp
Protocol  Address          Age (min)  Hardware Addr    Type   Interface
Internet  192.168.1.100       -       d0c7.89ab.11c0   ARPA   Vlan1
Internet  192.168.1.254       0       0007.b400.0103   ARPA   Vlan1
//以上表明 S1 的 ARP 请求获得网关（192.168.1.254）的 MAC 地址为 00-07-b4-00-01-03。
```

```
S1#clear arp-cache
S1#ping 4.4.4.4
S1#show ip arp
Protocol   Address          Age (min)   Hardware Addr     Type   Interface
Internet   192.168.1.100        -       d0c7.89ab.11c0    ARPA   Vlan1
Internet   192.168.1.254        0       0007.b400.0104    ARPA   Vlan1
```
//以上表明 S1 的 ARP 请求获得网关（192.168.1.254）的 MAC 地址为 **00-07-b4-00-01-04**。

也就是说在 GLBP 响应 ARP 请求时，每次会用不同的 MAC 地址响应，从而实现负载平衡。

（5）检查 GLBP 的冗余功能

首先在 S1 上用"**show ip arp**"命令确认 192.168.1.254 的 MAC 地址，从而确定出当前究竟是哪个路由器在实际转发数据。这里 192.168.1.254 的 MAC 地址为 00-07-b4-00-01-03（即 R3），从步骤（4）得知是 R3 在转发数据。

在 S1 上连续 ping 4.4.4.4，并在 R3 上关闭 G0/1 端口，观察 S1 的通信情况：

```
S1#ping 4.4.4.4 repeat 10000000
!!!!!!!!!!!!!!!!!!!!!!!!!!!!!!!!!!!!!!!!!!!!!!!!!!!!!!!!!!!!!!!!!!!!!!!!!!!!!!!!!!!!!!!!!!!!!!!!!!
!!!!!!!!!!!!!!!!!!!!!!!!!!!!!!!!!!U.U.U.U.U.U.U.U.U.U.U.U.U.U.U...................!!!!!!!!!!!!!!!!!!!
```

从以上可以看到当 R3 的端口 G0/1 发生故障后，其他路由器很快接替了它的工作，通信受到短暂的影响。因此 GLBP 不仅有负载平衡的能力，也有冗余的能力。可以使用"**show glbp**"命令在各路由器上查看一下谁是 00-07-b4-00-01-03 这个 MAC 地址的新的活动路由器。如下所示：

```
R4#show glbp
GigabitEthernet0/0 - Group 1
  State is Listen
  Virtual IP address is 192.168.1.254
  Hello time 3 sec, hold time 10 sec
    Next hello sent in 1.056 secs
  Redirect time 600 sec, forwarder timeout 14400 sec
  Authentication MD5, key-string
  Preemption enabled, min delay 1 sec
  Active is 192.168.1.1, priority 200 (expires in 10.880 sec)
  Standby is 192.168.1.2, priority 180 (expires in 8.640 sec)
  Priority 140 (configured)
  Weighting 200 (configured 200), thresholds: lower 170, upper 180
    Track object 100 state Up decrement 50
  Load balancing: round-robin
  IP redundancy name is "TEST"
  Group members:
    f872.ea69.1c78 (192.168.1.2) authenticated
    f872.eac8.4f98 (192.168.1.4) local
    f872.ead6.f4c8 (192.168.1.1) authenticated
  There are 4 forwarders (2 active)
  Forwarder 1
    State is Listen
```

MAC address is 0007.b400.0101 (learnt)

Owner ID is f872.ead6.f4c8

Time to live: 14399.616 sec (maximum 14400 sec)

Preemption enabled, min delay 25 sec

Active is 192.168.1.1 (primary), weighting 200 (expires in 9.824 sec)

Forwarder 2

State is Listen

2 state changes, last state change 00:01:05

MAC address is 0007.b400.0102 (learnt)

Owner ID is f872.ea69.1c78

Time to live: 14398.656 sec (maximum 14400 sec)

Preemption enabled, min delay 25 sec

Active is 192.168.1.2 (primary), weighting 200 (expires in 9.792 sec)

Forwarder 3   //R4 成为该 AVF 的活动路由器了，之前 R3 是活动路由器

State is Active

1 state change, last state change 00:00:04

MAC address is 0007.b400.0103 (learnt)

Owner ID is f872.ea69.18b8

Time to live: 14383.104 sec (maximum 14387 sec)

Preemption enabled, min delay 25 sec

Active is local, weighting 200

Forwarder 4

State is Active

1 state change, last state change 00:17:37

MAC address is 0007.b400.0104 (default)

Owner ID is f872.eac8.4f98

Preemption enabled, min delay 25 sec

Active is local, weighting 200

（6）Debug 信息

```
R1#debug glbp
*Feb  4 03:38:30.703: GLBP: Gi0/0 Grp 1 Hello   in  VG Standby pri 180 vIP 192.168.1.254 hello 3000, hold 10000 VF 2 Active   pri 167 vMAC 0007.b400.0102
*Feb  4 03:38:30.751: GLBP: Gi0/0 Grp 1 Hello   out VG Active  pri 200 vIP 192.168.1.254 hello 3000, hold 10000 VF 1 Active   pri 167 vMAC 0007.b400.0101
*Feb  4 03:38:31.511: GLBP: Gi0/0 Grp 1 Hello   in  VG Listen  pri 140 vIP 192.168.1.254 hello 3000, hold 10000 VF 3 Active   pri 135 vMAC 0007.b400.0103 VF 4 Active   pri 167 vMAC 0007.b400.0104
*Feb  4 03:38:35.647: GLBP: Gi0/0 Grp 1 Hello   out VG Active  pri 200 vIP 192.168.1.254 hello 3000, hold 10000 VF 1 Active   pri 167 vMAC 0007.b400.0101
*Feb  4 03:38:35.823: GLBP: Gi0/0 Grp 1 Hello   in  VG Standby pri 180 vIP 192.168.1.254 hello 3000, hold 10000 VF 2 Active   pri 167 vMAC 0007.b400.0102
*Feb  4 03:38:36.631: GLBP: Gi0/0 Grp 1 Hello   in  VG Listen  pri 140 vIP 192.168.1.254 hello 3000, hold 10000 VF 3 Active   pri 135 vMAC 0007.b400.0103 VF 4 Active   pri 167 vMAC 0007.b400.0104
```

```
*Feb    4 03:38:38.519: GLBP: Gi0/0 Grp 1 Hello    in    VG Standby pri 180 vIP 192.168.1.254 hello 3000,
hold 10000 VF 2 Active    pri 167 vMAC 0007.b400.0102
    *Feb    4 03:38:38.591: GLBP: Gi0/0 Grp 1 Hello    out VG Active   pri 200 vIP 192.168.1.254 hello 3000,
hold 10000 VF 1 Active    pri 167 vMAC 0007.b400.0101
    *Feb    4 03:38:39.607: GLBP: Gi0/0 Grp 1 Hello    in    VG Listen  pri 140 vIP 192.168.1.254 hello 3000,
hold 10000 VF 3 Active    pri 135 vMAC 0007.b400.0103 VF 4 Active    pri 167 vMAC 0007.b400.0104
    *Feb    4 03:38:41.311: GLBP: Gi0/0 Grp 1 Hello    out VG Active   pri 200 vIP 192.168.1.254 hello 3000,
hold 10000 VF 1 Active    pri 167 vMAC 0007.b400.0101
    *Feb    4 03:38:41.367: GLBP: Gi0/0 Grp 1 Hello    in    VG Standby pri 180 vIP 192.168.1.254 hello 3000,
hold 10000 VF 2 Active    pri 167 vMAC 0007.b400.0102
    *Feb    4 03:38:42.455: GLBP: Gi0/0 Grp 1 Hello    in    VG Listen  pri 140 vIP 192.168.1.254 hello 3000,
hold 10000 VF 3 Active    pri 135 vMAC 0007.b400.0103 VF 4 Active    pri 167 vMAC 0007.b400.0104
    *Feb    4 03:38:43.743: GLBP: Gi0/0 Grp 1 Hello    out VG Active   pri 200 vIP 192.168.1.254 hello 3000,
hold 10000 VF 1 Active    pri 167 vMAC 0007.b400.0101
    *Feb    4 03:38:44.247: GLBP: Gi0/0 Grp 1 Hello    in    VG Standby pri 180 vIP 192.168.1.254 hello 3000,
hold 10000 VF 2 Active    pri 167 vMAC 0007.b400.0102
```

5.5 实验 4：SLB

1. 实验目的

通过本实验，读者可以掌握：
① SLB 的工作原理。
② SLB 的配置。

2. 实验拓扑

SLB 实验拓扑如图 5-11 所示。由于 CISCO 2911 路由器不支持 SLB，因为本拓扑中的路由器 R2 用 CISCO 3640 路由器替换了原来的 CISCO 2911 路由器，IOS 使用的是"c3640-jk9o3s-mz.124-12.bin"。

R3 和 R4 作为 Telnet Server 使用，R1 作为测试用的计算机，在 R2 上配置 SLB。注意：本实验并不是使用图 1-1 所示拓扑来实现的。

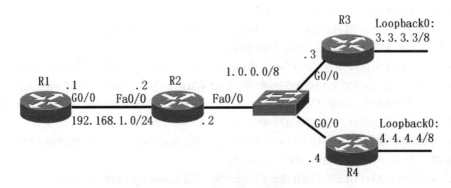

图 5-11 SLB 实验拓扑

提示

读者可以使用 GNS3 来完成该实验，GNS3 是一款具有图形化界面可以运行在多平台（包括 Windows, Linux, and MacOS 等）的网络虚拟软件，用于虚拟体验 Cisco 网际操作系统 IOS 或者是检验将要在真实的路由器上部署实施的相关配置。

3. 实验步骤

（1）配置 IP 地址、路由以及 Telnet 服务

```
R1(config)#interface GigabitEthernet0/0
R1(config-if)#no shutdown
R1(config-if)#ip address 192.168.1.1 255.255.255.0
R1(config)#ip route 0.0.0.0   0.0.0.0 192.168.1.2


R2(config)#interface FastEthernet0/0
R2(config-if)#no shutdown
R2(config-if)#ip address 192.168.1.2 255.255.255.0
R2(config)#interface FastEthernet1/0
R2(config-if)#no shutdown
R2(config-if)#ip address 1.1.1.2 255.0.0.0


R3(config)#interface GigabitEthernet0/0
R3(config-if)#no shutdown
R3(config-if)#ip address 1.1.1.3 255.0.0.0
R3(config-if)#exit
R3(config)#interface Loopback0
R3(config-if)#ip address 3.3.3.3 255.0.0.0
R3(config-if)#exit
R3(config)#line vty 0 4
R3(config-line)#password cisco
R3(config-line)#login
R3(config)#ip route 0.0.0.0   0.0.0.0 1.1.1.2


R4(config)#interface GigabitEthernet0/0
R4(config-if)#no shutdown
R4(config-if)#ip address 1.1.1.4 255.0.0.0
R4(config-if)#exit
R4(config)#interface Loopback0
R4(config-if)#ip address 4.4.4.4 255.0.0.0
R4(config-if)#exit
R4(config)#line vty 0 4
R4(config-line)#password cisco
R4(config-line)#login
R4(config)#ip route 0.0.0.0   0.0.0.0   1.1.1.2
```

（2）配置分派模式的 SLB

R2(config)#**ip slb serverfarm TELNET-SERVER**
//创建服务器组

R2(config-slb-sfarm)#**real 1.1.1.3**
//其中的一个服务器是 1.1.1.3，即 R3

R2(config-slb-real)#**weight 1**
//配置权重，SLB 根据各服务器的权重计算承担的负载大小

R2(config-slb-real)#**inservice**
//启用服务器

R2(config-slb-real)#**faildetect**
//配置将检测服务器的可用状态，如果服务器发生故障，SLB 将不会把流量转发到该服务器

R2(config-slb-sfarm)#**real 1.1.1.4**
//另一个服务器是 1.1.1.4，即 R4

R2(config-slb-real)#**weight 1**

R2(config-slb-real)#**inservice**

R2(config-slb-real)#**faildetect**

R2(config-slb-real)#**exit**

R2(config-slb-sfarm)#**predictor roundrobin**
//配置采用轮询进行负载均衡，也可以选择 "leastconns"，则由连接数最少的服务器承担负载

 提示

默认时，SLB 采用的是分派模式，该模式要求运行 SLB 的路由器和真实的服务器必须在同一子网中，二层可达。如图 5-11 中的 R2 及 R3 和 R4，它们都同在 1.0.0.0/255.0.0.0 网络中，并且是二层邻居。

R2(config)#**ip slb vserver V-TELNET-SERVER**
//创建虚拟的服务器

R2(config-slb-vserver)# **virtual 1.1.1.100 tcp telnet**
//虚拟出的是 Telnet 服务器，IP 地址为 1.1.1.100，客户端计算机通过该 IP 地址来访问真实的 Telnet 服务器。在分派模式中，该地址要和真实的服务器在同一子网中

R2(config-slb-vserver)# **serverfarm TELNET-SERVER**
//使用前面创建的服务器组

R2(config-slb-vserver)# **client 192.168.1.0 255.255.255.0**
//限定客户计算机的 IP 地址

R2(config-slb-vserver)# **inservice**
//启用虚拟的服务器

R3(config)#**int GigabitEthernet0/0**

R3(config-if)#**ip address 1.1.1.100 255.0.0.0 secondary**
//在真实的 Telnet 服务器上增加一个地址作为第二地址，该地址必须是虚拟服务器的 IP 地址。这是因为 R2（SLB 服务器）会把目的 IP 地址为 1.1.1.100 的数据包分派到真实的服务器上，如果该服务器没有 1.1.1.100 这个地址，服务器将会丢弃数据包

R4(config)#**int GigabitEthernet0/0**

R4(config-if)#**ip address 1.1.1.100 255.0.0.0 secondary**

//R4 上也配置了相同的 IP 地址。

 提示

当 R3、R4 采用 Secondary IP 增加 IP 地址时,R3 和 R4 会报 "IP 地址重复" 的错误,虽然这不影响使用。

(3) 测试和检查 SLB

从 R1 上反复 Telnet 1.1.1.100,看看是否会实现负载均衡。如下所示:

```
R1#telnet 1.1.1.100
Trying 1.1.1.100 ... Open
User Access Verification
Password:
R4>exit
[Connection to 1.1.1.100 closed by foreign host]
//可以看到连接到了 R4

R1#telnet 1.1.1.100
Trying 1.1.1.100 ... Open
User Access Verification
Password:
R3>exit
[Connection to 1.1.1.100 closed by foreign host]
//可以看到连接到了 R3,证实负载均衡在起作用

R2#show ip slb conns
vserver           prot client                    real            state      nat
-----------------------------------------------------------------------------------
V-TELNET-SERVER   TCP  192.168.1.1:23928         1.1.1.4         ESTAB      none
//可以看到一个客户连接到 1.1.1.4 这一真实服务器上

R2#show ip slb conns detail
V-TELNET-SERVER, client = 192.168.1.1:23928
    state= ESTAB, real = 1.1.1.4, nat= none
    v_ip= 1.1.1.100:23, TCP, service = none
    client_syns = 1, sticky = FALSE, flows attached = 0
//以上是连接的详细信息

R2#show ip slb serverfarms
server farm            predictor         nat     reals    bind id
-----------------------------------------------------------------------
TELNET-SERVER          ROUNDROBIN        none    2        0
//以上是服务器组的基本信息,采用的是 ROUNDROBIN 负载均衡方式,有 2 个真实的服务器

R2#show ip slb serverfarms detail
```

TELNET-SERVER, predictor = ROUNDROBIN, nat = none
      virtuals inservice: 1, reals = 2, bind id = 0
  Real servers:
    1.1.1.3, weight = 1, OPERATIONAL, conns = 0   //真实服务器的 IP 地址，服务器的状态是可用
    1.1.1.4, weight = 1, OPERATIONAL, conns = 0
  Total connections = 0

R2#show ip slb vservers

slb vserver	prot	virtual	state	conns
V-TELNET-SERVER	TCP	1.1.1.100:23	INSERVICE	0

//以上显示虚拟服务器的基本信息

R2#show ip slb vservers detail
V-TELNET-SERVER, state = INSERVICE, v_index = 3
  virtual = 1.1.1.100:23, TCP, service = none, advertise = TRUE
  server farm = TELNET-SERVER, delay = 10, idle = 3600
  sticky timer = 0, sticky subnet = 255.255.255.255
  sticky group id = 0
  synguard counter = 0, synguard period = 0
  conns = 0, syns = 7, syn drops = 0
  standby group = none
  Clients:
  192.168.1.0   255.255.255.0

R2#show ip slb stats
Pkts via normal switching: 215
Pkts via special switching: 0
Connections Created: 5
Connections Established: 5
Connections Destroyed: 5
Connections Reassigned: 0
Zombie Count: 0
//以上显示 SLB 的统计数

（4）配置定向模式的 SLB

R2(config)#**ip slb serverfarm ANOTHER**
R2(config-slb-sfarm)# **nat server**
//配置 SLB 的模式为定向，默认是分派模式
R2(config-slb-sfarm)#**predictor roundrobin**
R2(config-slb-sfarm)# **real 3.3.3.3**
//以上是真实服务器的 IP 地址，这里使用 R3 的环回口作为 Telnet 服务器的地址
R2(config-slb-real)# **weight 1**

```
R2(config-slb-real)#    faildetect
R2(config-slb-real)#    inservice
R2(config-slb-real)#    real 4.4.4.4
R2(config-slb-real)#    weight 1
R2(config-slb-real)#    faildetect
R2(config-slb-real)#    inservice

R2(config-slb-real)#ip slb vserver V-ANOTHER
R2(config-slb-vserver)# virtual 172.16.1.100 tcp telnet
//在定向模式中，虚拟出的 IP 地址和真实服务器的 IP 地址并不需要在同一子网中
R2(config-slb-vserver)# serverfarm ANOTHER
R2(config-slb-vserver)# client 192.168.1.0 255.255.255.0
R2(config-slb-vserver)# inservice


R2(config)#ip route 3.0.0.0 255.0.0.0 1.1.1.3
R2(config)#ip route 4.0.0.0 255.0.0.0 1.1.1.4
```

//在定向模式中，路由器会更改客户计算机发来的数据包，把目的 IP 地址 172.16.1.100 改为真实服务器的 IP 地址（3.3.3.3 或者 4.4.4.4）。采用以上命令添加路由，保证 R2 能把数据包发到 R3 或者 R4 上。

同样，从 R1 反复 Telnet 172.16.1.100，看看是否会实现负载均衡。如下所示：

```
R1#telnet 172.16.1.100
Trying 172.16.1.100 ... Open
User Access Verification
Password:
R3>exit
[Connection to 172.16.1.100 closed by foreign host]


R1#telnet 172.16.1.100
Trying 172.16.1.100 ... Open
User Access Verification
Password:
R4>exit
[Connection to 172.16.1.100 closed by foreign host]
```
//可以证实负载均衡在起作用

```
R2#show ip slb serverfarms detail
ANOTHER, predictor = ROUNDROBIN, nat = SERVER    //采用的是定向模式
        virtuals inservice: 1, reals = 2, bind id = 0
  Real servers:
      3.3.3.3, weight = 1, OPERATIONAL, conns = 0
      4.4.4.4, weight = 1, OPERATIONAL, conns = 0
  Total connections = 0
TELNET-SERVER, predictor = ROUNDROBIN, nat = none    //采用的是分派模式
        virtuals inservice: 1, reals = 2, bind id = 0
```

Real servers:
   1.1.1.3, weight = 1, OPERATIONAL, conns = 0
   1.1.1.4, weight = 1, OPERATIONAL, conns = 0
Total connections = 0

R2#**show ip slb conns**

vserver	prot client	real	state	nat
V-ANOTHER	TCP 192.168.1.1:43704	3.3.3.3	ESTAB	S

提示

在配置 SLB 时，真实服务器上的 IP 地址和路由是经常被忽视的地方，请根据分派或在定向模式中 SLB 路由器发送到真实服务器的数据包中的目的 IP 地址，以及真实服务器的回包中的目的 IP 地址，核实 SLB 路由器和真实的服务器是否存在相应的路由表。

5.6 实验 5：Syslog

1. 实验目的

通过本实验可以掌握：
① 日志服务器软件的使用方法。
② 把路由器或者交换机日志发送到服务器的配置。

2. 实验拓扑

Syslog 实验拓扑如图 5-12 所示。

图 5-12　Syslog 实验拓扑

3. 实验步骤

（1）配置交换机

S1(config)#**logging on**　　//打开交换机的日志功能，默认就是打开的
S1(config)#**logging console debugging**
//把日志在控制台上显示出来，默认就是这样，所以我们能在控制台上看到各种信息。这里还指明了显示的是比 debugging 等级（等级 7）还高的信息，debugging 等级（等级 7）是最低的级别
S1(config)#**logging monitor debugging**
　　//把日志在远程用户的终端上显示，用户还得在终端上的特权模式下执行"**terminal monitor**"命令才行

S1(config)#**logging buffer debugging**
//把日志记录在内存中，以后使用"**show logging**"命令就可以看到日志了
S1(config)#**logging host 172.16.1.100**
S1(config)#**logging trap debugging**
//配置交换机把日志发送到专门的日志服务器，服务器地址为 172.16.1.100

S1(config)#**logging origin-id ip**
//在指明交换机发送日志时，会用 IP 地址作为 ID，当然也可以用主机名
S1(config)#**logging facility local7**   //将记录事件类型定义为 local7
S1(config)#**logging source-interface vlan 1**
//指明交换机以 VLAN1 端口的 IP 地址作为源 IP 地址向服务器发送日志
S1(config)#**service timestamps log**   //指明日志中要加上发生时间
S1(config)#**service timestamps log datetime**
//指明日志中的发生时间采用绝对时间（年、月、日、时、分、秒、毫秒），当然也可以采用交换机的开机时间（开了多长时间）
S1(config)#**service sequence-numbers**   //指明日志中要加入序号

可以查看日志的配置情况，如下所示：

S1#**show logging**
Syslog logging: enabled (0 messages dropped, 0 messages rate-limited, 0 flushes, 0 overruns, xml disabled, filtering disabled)
//以上是发送到 Syslog 服务器上的日志情况
No Active Message Discriminator.

No Inactive Message Discriminator.
    Console logging: level debugging, 32 messages logged, xml disabled,
                     filtering disabled
//以上是发送到控制台的日志情况
    Monitor logging: level debugging, 0 messages logged, xml disabled,
                     filtering disabled
//以上是发送到终端的日志情况
    Buffer logging: level debugging, 33 messages logged, xml disabled,
                     filtering disabled
//以上是发送到内存的日志情况
    Logging Exception size (4096 bytes)
    Count and timestamp logging messages: disabled
    Persistent logging: disabled
No active filter modules.
    Trap logging: level debugging, 37 message lines logged
        Logging to 172.16.1.100   (udp port 514, audit disabled,
            link up),
            3 message lines logged,
            0 message lines rate-limited,
            0 message lines dropped-by-MD,

xml disabled, sequence number disabled
filtering disabled
Logging Source-Interface: VRF Name:
Vlan1
Log Buffer (4096 bytes):
*Mar 1 00:01:17.141: %LINK-3-UPDOWN: Interface FastEthernet0/2, changed state to up
*Mar 1 00:01:17.267: %LINK-3-UPDOWN: Interface FastEthernet0/4, changed state to up
*Mar 1 00:01:28.181: %LINK-3-UPDOWN: Interface FastEthernet0/24, changed state to up
*Mar 1 00:01:28.222: %LINK-3-UPDOWN: Interface FastEthernet0/23, changed state to up
*Mar 1 00:01:29.187: %LINEPROTO-5-UPDOWN: Line protocol on Interface FastEthernet0/24, changed state to up
*Mar 1 00:01:29.229: %LINEPROTO-5-UPDOWN: Line protocol on Interface FastEthernet0/23, changed state to up
*Mar 1 00:01:34.581: %LINEPROTO-5-UPDOWN: Line protocol on Interface FastEthernet0/2, changed state to down
*Mar 1 00:01:35.512: %LINK-3-UPDOWN: Interface FastEthernet0/2, changed state to down
*Mar 1 00:01:45.050: %LINEPROTO-5-UPDOWN: Line protocol on Interface Vlan1, changed state to up
*Mar 1 00:18:34.753: %LINEPROTO-5-UPDOWN: Line protocol on Interface FastEthernet0/3, changed state to down
*Mar 1 00:18:35.760: %LINK-3-UPDOWN: Interface FastEthernet0/3, changed state to down
*Mar 1 00:19:46.946: %LINK-3-UPDOWN: Interface FastEthernet0/3, changed state to up
*Mar 1 00:19:47.952: %LINEPROTO-5-UPDOWN: Line protocol on Interface FastEthernet0/3, changed state to up
*Mar 1 00:36:05.795: %LINK-5-CHANGED: Interface Vlan1, changed state to administratively down
*Mar 1 00:36:05.804: %LINEPROTO-5-UPDOWN: Line protocol on Interface Vlan1, changed state to down
*Mar 1 00:36:33.897: %SYS-6-LOGGINGHOST_STARTSTOP: Logging to host 172.16.1.100 port 0 CLI Request Triggered
*Mar 1 00:36:34.904: %SYS-6-LOGGINGHOST_STARTSTOP: Logging to host 172.16.1.100 port 514 started - CLI initiated
000032: *Mar 1 00:37:24: %SYS-5-CONFIG_I: Configured from console by console
000033: *Mar 1 00:37:30: %SYS-6-LOGGINGHOST_STARTSTOP: Logging to host 172.16.1.100 port 514 started – reconnection
//以上段落是内存中保留的日志

（2）安装日志服务器

有各种各样的日志服务器软件，它们有的运行在 Windows 上，有的运行在 Unix/Linux 系列的主机上，商用的软件多数要和数据库连接。这里仅仅为了说明问题，选择了很简单的免费服务器，可以从 http://www.onlinedown.net/soft/54636.htm 下载，解压后直接运行即可。日志服务器如图 5-13 所示。

第 5 章　高可用性

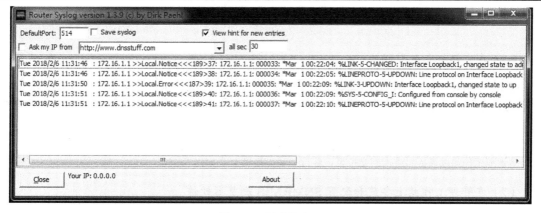

图 5-13　日志服务器

4．实验调试

在交换机 S1 上，打开和关闭一个端口就会产生日志，以下是其中的几条日志：

000038: *Mar   1 05:44:40: %LINEPROTO-5-UPDOWN: Line protocol on Interface Loopback100, changed state to up
000039: *Mar   1 05:44:52: %LINK-5-CHANGED: Interface Loopback100, changed state to administratively down
000040: *Mar   1 05:44:53: %LINEPROTO-5-UPDOWN: Line protocol on Interface Loopback100, changed state to down
000041: *Mar   1 05:44:54: %SYS-5-CONFIG_I: Configured from console by console
000042: *Mar   1 05:44:56: %LINK-3-UPDOWN: Interface Loopback100, changed state to up
000043: *Mar   1 05:44:57: %LINEPROTO-5-UPDOWN: Line protocol on Interface Loopback100, changed state to up

5.7　实验 6：SNMP

1．实验目的

通过本实验可以掌握：
① SNMP 的工作原理。
② 路由器或者交换机上的 SNMP 配置。

2．实验拓扑

SNMP 实验拓扑如图 5-14 所示。

图 5-14　SNMP 实验拓扑

3. 实验步骤

（1）在交换机上配置团体

S1(config)#**snmp-server community TEST-RO ro**
//配置团体名字。实际上可以把该团体看成密码，管理工作站以该密码连接到交换机，只能读取交换机上的 MIB 信息

S1(config)#**snmp-server community TEST-RW rw**
//管理工作站以该密码连接到交换机，能读/写交换机上的 MIB 信息

（2）在管理工作站上安装和配置 SNMP MIB 浏览器软件

可以从网站 https://sourceforge.net/projects/snmpb/?source=directory 免费下载 SnmpB 软件，SnmpB 软件是一款读取和更改 SNMP 代理中的 MIB 信息的软件。安装 SnmpB 后，打开该软件，SnmpB 程序界面如图 5-15 所示。

单击"Modules"选项卡，可以选择从 SNMP 代理（交换机）上要读取的 MIB 库。设置 SnmpB 加载的 MIB 模块如图 5-16 所示。

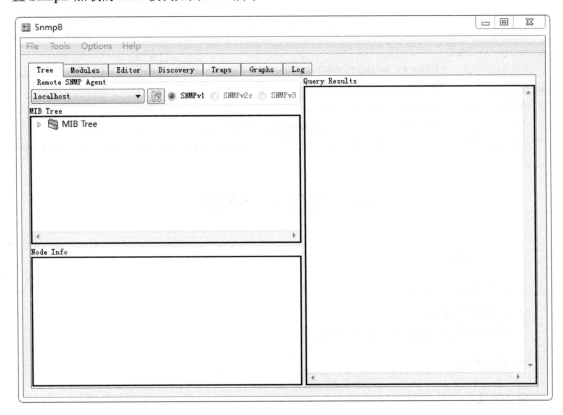

图 5-15　SnmpB 程序界面

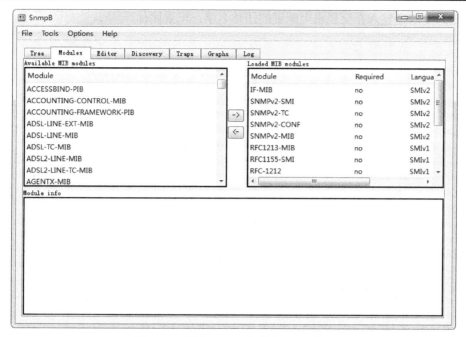

图 5-16　设置 SnmpB 加载的 MIB 模块

技术要点

该软件只支持通用的 MIB，但是不同厂家的设备有自己定义的 MIB，可以从"options"→"Preferences"→"Modules"→"Add"菜单，把各厂家的 MIB 加到软件中。Cisco 的 MIB 可以从 http://tools.cisco.com/ITDIT/MIBS/MainServlet 下载，然而遗憾的是下载的是".my"文件，而不是标准格式".txt"文件，需要使用 Ciscoworks 软件进行转换。因此建议使用商用的、能够支持".my"文件的 SNMP MIB 浏览软件，例如：HiliSoft MIB Browser。

从"Options"→"Manage Agent Profiles"菜单打开如图 5-17 所示窗口设置 Agent 的信息。在"Name"处输入一个名字，该名字仅供管理员使用。在"Agent Address/Name"处输入交换机的 IP 地址，选择使用 SNMP V1。

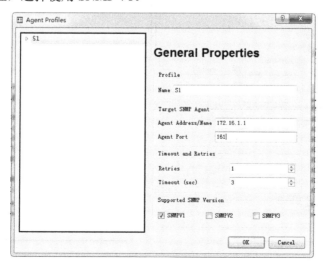

图 5-17　设置 Agent 的信息

如图 5-18 所示输入在交换机上配置的团体名字，配置团体。

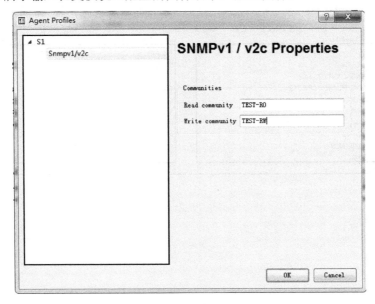

图 5-18　配置团体

（3）在管理工作站上读取和修改交换机信息

如图 5-19 所示展开 MIB 树，右键单击 system 项，选择"Walk"选项，则 SnmpB 会遍历该项下的 MIB 树。

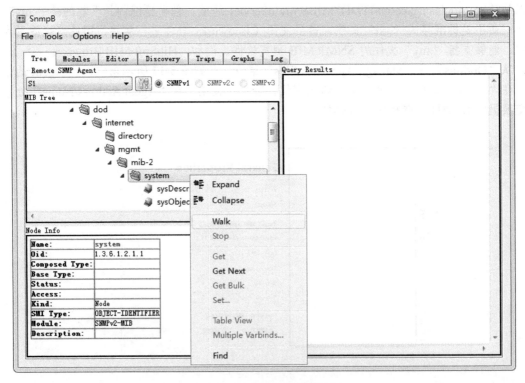

图 5-19　遍历 MIB 树

如图 5-19 所示，选择 sysName，右键单击它，选择"Set"选项，打开如图 5-20 所示窗口，设置 MIB 的值。输入"HELLO"，单击"OK"按钮，则发现交换机的 hostname 被修改为"HELLO"了（原来为 S1）。

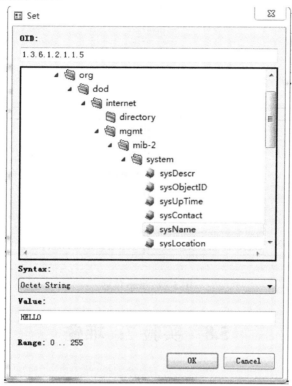

图 5-20　设置 MIB 的值

（4）在交换机上配置 SNMP trap

```
S1(config)#snmp-server host 172.16.1.100 traps S1
//以上指明管理工作站的 IP 地址，并且以名为 S1 的团体发送 trap 信息
S1(config)#snmp-server enable traps
//开启 trap


S1(config)#snmp-server contact Jack.Chen
//以上配置联系信息，可选


S1(config)#int loopback 0
S1(config-if)#shutdown
S1(config-if)#no shutdown
```

以上是人为制造的一些事件，如图 5-21 所示，会看到交换机发送到管理工作站的 trap 信息。

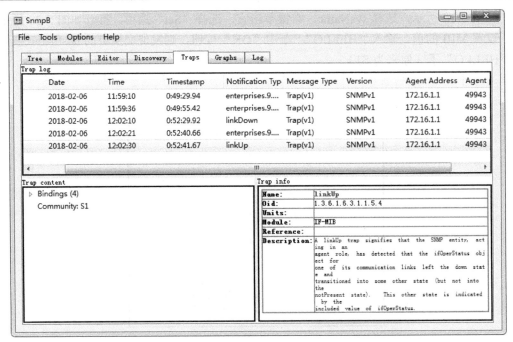

图 5-21 trap 信息

5.8 实验 7：堆叠

1. 实验目的

通过本实验可以掌握：
① 堆叠的工作原理。
② 堆叠的配置。

2. 实验拓扑

堆叠实验拓扑如图 5-22 所示，这里的交换机 S1 和 S2 采用的并不是贯穿本书的那个实验拓扑（见第 1 章图 1-1）中的交换机，这里的 S1 和 S2 是 Catalyst 3750 交换机，S1 的型号是 ws-c3750-24p, S2 的型号是 ws-c3750-24ts 。

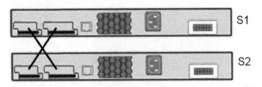

图 5-22 堆叠实验拓扑

3. 实验步骤

（1）准备工作

先把两交换机之间的堆叠线拆除，把两交换机的 IOS 更新为相同的 IOS。本书使用的 IOS

如下所示：

```
Switch#show flash
Directory of flash:/
2    -rwx    10737954    Mar 1 1993 00:10:18 +00:00    c3750-advipservicesk9-mz.122-44.SE.bin
```

确认两交换机上堆叠协议的版本号是相同的，如下所示：

```
Switch #show platform stack-manager all
(省略部分输出)
```

Stack State Machine View

Switch Number	Master/Member	Mac Address	Version (maj.min)	Current State
1	Master	001a.6ccf.bb80	1.37	Ready //堆叠协议版本为 1.37

(省略部分输出)

以下修改堆叠交换机的优先级，这里控制 S1 是 Master，S2 是 Member。

先确定交换机当前的编号，使用"**show running-config**"命令。

```
S1#show running-config
(省略部分输出)
no aaa new-model
switch 1 provision ws-c3750-24p    //1 就是当前交换机的编号，默认值都为 1，但有可能被更改过
system mtu routing 1500
ip subnet-zero
 (省略部分输出)
//以上说明当前交换机的编号为 1，因此下面采用编号 1 来修改交换机的优先级
S1(config)#switch 1 priority 12
Changing the Switch Priority of Switch Number 1 to 12
Do you want to continue?[confirm]
New Priority has been set successfully
//用同样的方式把 S2 的优先级改为 10
S2(config)#switch 1 priority 10
Changing the Switch Priority of Switch Number 1 to 10
Do you want to continue?[confirm]
New Priority has been set successfully
//优先级数值大的交换机会成为 Master
```

技术要点

默认时全部交换机的编号都为 1。堆叠后交换机开机，如果有编号相同的交换机存在，则交换机将使用最小的、空闲的编号，编号会发生变化。一旦编号确定，即使将交换机从堆叠中脱离出来，编号也不会发生变化。因此如果有人使用过某交换机做堆叠，使用"**show running-config**"命令可能看到不是 1 的交换机编号，可以使用稍后介绍的"**switch 2 renumber 1**"命令修改交换机的编号。

（2）堆叠交换机

如图 5-22 所示，当关闭交换机电源后，连接好堆叠线，把两个交换机开机。交换机开机后，检测到有堆叠后，会等待一段时间让其他交换机也完成开机，等待完毕后，选举出 Master，一旦 Master 选出，Master 不会被其他后开机的高优先级的交换机抢占。如下所示：

```
(省略部分输出)
OST: PortASIC RingLoopback Tests : Begin
POST: PortASIC RingLoopback Tests : End, Status Passed
SM: Detected stack cables at PORT1 PORT2        //检测到有堆叠
Waiting for Stack Master Election...             //等待 Master 选举
SM: Waiting for other switches in stack to boot... //等待其他交换机开机
###
SM: All possible switches in stack are booted up

POST: Inline Power Controller Tests : Begin
POST: Inline Power Controller Tests : End, Status Passed

POST: PortASIC CAM Subsystem Tests : Begin
POST: PortASIC CAM Subsystem Tests : End, Status Passed

POST: PortASIC Stack Port Loopback Tests : Begin   //堆叠测试
POST: PortASIC Stack Port Loopback Tests : End, Status Passed

POST: PortASIC Port Loopback Tests : Begin
POST: PortASIC Port Loopback Tests : End, Status Passed

Election Complete                //选举完毕
Switch 1 booting as Master       //交换机 S1 的优先级高，成为 Master
Waiting for Port download...Complete

(省略部分输出)
cisco WS-C3750-24P (PowerPC405) processor (revision J0) with 118784K/12280K bytes of memory.
Processor board ID CAT1048NG92
Last reset from power-on
1 Virtual Ethernet interface
48 FastEthernet interfaces
4 Gigabit Ethernet interfaces
The password-recovery mechanism is enabled.

512K bytes of flash-simulated non-volatile configuration memory.
Base ethernet MAC Address       : 00:1A:6C:CF:BB:80
Motherboard assembly number     : 73-9672-09
Power supply part number        : 341-0029-05
Motherboard serial number       : CAT10475S31
```

```
Power supply serial number        : DTH1045C1FK
Model revision number             : J0
Motherboard revision number       : A0
Model number                      : WS-C3750-24PS-S
System serial number              : CAT1048NG92
Top Assembly Part Number          : 800-25860-04
Top Assembly Revision Number      : B0
Version ID                        : V05
CLEI Code Number                  : CNMV1K0CRD
Hardware Board Revision Number    : 0x01


Switch Ports Model               SW Version          SW Image
------ ----- ------------        ----------          ----------
*    1 26    WS-C3750-24P        12.2(44)SE          C3750-ADVIPSERVICESK9-M
     2 26    WS-C3750-24TS       12.2(44)SE          C3750-ADVIPSERVICESK9-M
//以上可以看到交换机 S1 的编号为 1，交换机 S2 的编号为 2


Switch 02    //在 Master 上（交换机 S1 上），可以看到另一交换机的信息
---------
Switch Uptime                     : Unknown
Base ethernet MAC Address         : 00:16:C7:62:E9:80
Motherboard assembly number       : 73-9677-09
Power supply part number          : 341-0034-01
Motherboard serial number         : CAT10010B8P
Power supply serial number        : DAB09510RFC
Model revision number             : L0
Motherboard revision number       : A0
Model number                      : WS-C3750-24TS-E
System serial number              : CAT0910N247
Top assembly part number          : 800-25857-02
Top assembly revision number      : C0
Version ID                        : V05
CLEI Code Number                  : CNMV100CRE


Press RETURN to get started!

//以下是堆叠的启动情况

*Mar  1 00:00:40.835: %STACKMGR-4-SWITCH_ADDED: Switch 1 has been ADDED to the stack
*Mar  1 00:00:40.835: %STACKMGR-4-SWITCH_ADDED: Switch 2 has been ADDED to the stack
*Mar  1 00:00:55.968: %LINEPROTO-5-UPDOWN: Line protocol on Interface Vlan1, changed state to down
*Mar  1 00:00:57.319: %SPANTREE-5-EXTENDED_SYSID: Extended SysId enabled for type vlan
*Mar  1 00:01:01.161: %SYS-5-CONFIG_I: Configured from memory by console
*Mar  1 00:01:01.220: %STACKMGR-5-SWITCH_READY: Switch 1 is READY
```

*Mar　1 00:01:01.220: %STACKMGR-4-STACK_LINK_CHANGE: Stack Port 1 Switch 1 has changed to state UP

　　*Mar　1 00:01:01.220: %STACKMGR-4-STACK_LINK_CHANGE: Stack Port 2 Switch 1 has changed to state UP

　　*Mar　1 00:01:01.555: %STACKMGR-5-MASTER_READY: Master Switch 1 is READY

　　*Mar　1 00:01:01.949: %SYS-5-RESTART: System restarted --

　　Cisco IOS Software, C3750 Software (C3750-ADVIPSERVICESK9-M), Version 12.2(44)SE, RELEASE SOFTWARE (fc1)

　　Copyright (c) 1986-2008 by Cisco Systems, Inc.

　　Compiled Sat 05-Jan-08 00:29 by weiliu

　　*Mar　1 00:01:01.966: %SSH-5-ENABLED: SSH 1.99 has been enabled

　　*Mar　1 00:01:06.991: %STACKMGR-5-SWITCH_READY: Switch 2 is READY

　　*Mar　1 00:01:06.991: %STACKMGR-4-STACK_LINK_CHANGE: Stack Port 1 Switch 2 has changed to state UP

　　*Mar　1 00:01:06.991: %STACKMGR-4-STACK_LINK_CHANGE: Stack Port 2 Switch 2 has changed to state UP

　　*Mar　1 00:00:35.408: %STACKMGR-4-SWITCH_ADDED: Switch 1 has been ADDED to the stack (Switch-2)

　　*Mar　1 00:00:35.408: %STACKMGR-4-SWITCH_ADDED: Switch 2 has been ADDED to the stack (Switch-2)

　　*Mar　1 00:01:04.600: %SPANTREE-5-EXTENDED_SYSID: Extended SysId enabled for type vlan (Switch-2)

　　*Mar　1 00:01:06.899: %SYS-5-CONFIG_I: Configured from memory by console (Switch-2)

　　*Mar　1 00:01:06.924: %STACKMGR-5-SWITCH_READY: Switch 1 is READY (Switch-2)

　　*Mar　1 00:01:07.511: %STACKMGR-5-MASTER_READY: Master Switch 1 is READY (Switch-2)

　　*Mar　1 00:01:07.654: %SYS-5-RESTART: System restarted -- (Switch-2)

　　Cisco IOS Software, C3750 Software (C3750-ADVIPSERVICESK9-M), Version 12.2(44)SE, RELEASE SOFTWARE (fc1) (Switch-2)

　　Copyright (c) 1986-2008 by Cisco Systems, Inc. (Switch-2)

　　Compiled Sat 05-Jan-08 00:29 by weiliu (Switch-2)

　　*Mar　1 00:01:07.662: %STACKMGR-5-SWITCH_READY: Switch 2 is READY (Switch-2)

　　*Mar　1 00:01:10.388: %LINEPROTO-5-UPDOWN: Line protocol on Interface FastEthernet2/0/5, changed state to up

　　*Mar　1 00:01:11.764: %LINK-3-UPDOWN: Interface FastEthernet2/0/5, changed state to up

　　*Mar　1 00:01:40.789: %LINEPROTO-5-UPDOWN: Line protocol on Interface Vlan1, changed state to up

提示

　　Master 不具有抢占性，因此如果某一交换机先开机并且成为 Master 后，其他更高优先级的交换机后开机，它也只能是 Member 交换机。因此，也可以通过控制开机时间来控制 Master 的选举。

　　（3）检查堆叠

　　堆叠后两台交换机形成一台逻辑上的交换机，可以在任何一台交换机上执行命令，效果是一样的。两台交换机的端口将统一编号，编号为 1 的交换机端口编号形式为 interface

FastEthernet **1**/0/1,而编号为 2 的交换机端口编号形式为 interface FastEthernet **2**/0/1。

```
Switch#show switch
Switch/Stack Mac Address : 001a.6ccf.bb80
                                           H/W       Current
Switch#   Role      Mac Address    Priority Version  State
--------------------------------------------------------------
*1        Master 001a.6ccf.bb80    12       0        Ready
 2        Member 0016.c762.e980    10       0        Ready
```

各字段含义如下所述。

① Switch#:交换机编号。

② Role:角色,有 Master 或者 Member。

③ Mac Address:交换机的 MAC 地址。

④ Priority:优先级。

⑤ H/W Version:硬件版本。

⑥ Current State:当前状态,如果不是 Ready,则大多是堆叠交换机的硬件型号或者 IOS 不匹配。

以下显示交换机的邻居:

```
Switch#show switch neighbors
  Switch #    Port 1      Port 2
  --------    ------      ------
     1           2           2
//表示编号为 1 的交换机的 Port1 和 Port2 都接到编号为 2 的交换机上
     2           1           1
```

以下显示各交换机上堆叠口的状态:

```
Switch#show switch stack-ports
  Switch #    Port 1      Port 2
  --------    ------      ------
     1          Ok          Ok
     2          Ok          Ok
```

以下显示堆叠环的速度:

```
Switch#show switch stack-ring speed
Stack Ring Speed         : 32G         //32Gbps 的速度
Stack Ring Configuration : Full        //全带宽功能
Stack Ring Protocol      : StackWise   //堆叠协议
```

以下显示堆叠环已传送的帧数量,不包含 SPAN 和一些特别的帧数量:

```
Switch#show switch stack-ring activity
  Sw    Frames sent to stack ring (approximate)
  ------------------------------------------
  1        54610    //编号为 1 的交换机发送的帧数量
  2        56823

Total frames sent to stack ring : 111433    //帧总数
```

Note: these counts do not include frames sent to the ring by certain output features, such as output SPAN and output ACLs.

以下显示堆叠后交换机的配置：

Switch#**show running-config**
Building configuration...

Current configuration : 3860 bytes
!
version 12.2
no service pad
service timestamps debug datetime msec
service timestamps log datetime msec
no service password-encryption
!
hostname Switch
!
boot-start-marker
boot-end-marker
!
!
no aaa new-model
switch 1 provision ws-c3750-24p   //可以看到交换机的编号和硬件型号
switch 2 provision ws-c3750-24ts   //可以看到交换机的编号和硬件型号
(省略部分输出)

（4）拆除堆叠

关闭电源，拆除堆叠线，即使交换机已经脱离堆叠，交换机的编号也不会改变，因此交换机 S2 的编号仍为 2，它的端口编号形式仍为 interface FastEthernet 2/0/1，可以使用命令把编号回复为 1，如下所示（在 S2 交换机上进行）：

Switch(config)#**switch 2 renumber 1**
WARNING: Changing the switch number may result in a
configuration change for that switch.
The interface configuration associated with the old switch
number will remain as a provisioned configuration.
Do you want to continue?[confirm]
Changing Switch Number 2 to Switch Number 1   //编号已修改
New Switch Number will be effective after next reboot //重启交换机才能生效

5.9 本章小结

本章介绍了 HSRP 和 VRRP 的目的和基本工作原理。HSRP 和 VRRP 都是为了实现网关的冗余，它们把多个路由器组成一个小组，选出活动路由器，当活动路由器出现故障时，其他路由器接替它的工作。GLBP 则不仅具有网络冗余功能，还可以提供负载平衡的功能。SLB 是用于服务器负载均衡的，可以在路由器上实现负载均衡，而不需在服务器上采用群集软件。以上这些技术都可以实现设备的冗余性，本章详细介绍了它们的配置。此外，为了增加网络设备的可用性，可以使用日志服务器或者网络管理软件对设备进行监控，常用的有 Syslog 日志服务器和 SNMP 工作站。

第 6 章 交换机的安全

交换机作为局域网中最常见的设备，在安全上面临着重大威胁。这些威胁有的是针对交换机管理上的漏洞，攻击者试图控制交换机；有的针对的是交换机的功能，攻击者试图扰乱交换机的正常工作，从而达到破坏甚至窃取数据的目的。本章将介绍一些减少交换机可能被攻击的手段。

6.1 交换机的安全简介

针对交换机的攻击有以下几类：
① 交换机配置／管理的攻击。
② MAC 泛洪攻击。
③ DHCP 欺骗攻击。
④ MAC 泛洪和 IP 地址欺骗攻击。
⑤ ARP 欺骗。
⑥ VLAN 跳跃攻击。
⑦ STP 攻击。
⑧ VTP 攻击。

表 6-1 各种攻击类型及其缓解措施

攻击方法	描述	缓解措施
CDP 和 LLDP	CDP 和 LLDP 发送设备本身的信息	除设备互联的端口外，关闭 CDP 和 LLDP
Telnet	Telnet 用于远程管理设备，以明文发送密码，容易被窃听	使用 SSH 和用 ACL 限制客户端的 IP 地址
各种网络服务，如 finger	为了方便，默认时网络设备开启了很多常用服务	保留有必要的服务，关闭不必要的服务
MAC 泛洪攻击	通过发送虚假源 MAC 地址的帧，占满 MAC 地址表，导致交换机泛洪数据帧	启用端口安全
DHCP 欺骗攻击	先冒充 DHCP Client 申请 IP 地址，耗尽 DHCP Server 的 IP 地址，然后再冒充 DHCP Server 分配 IP 地址	启用 DHCP Snooping，这是 DAI 和 IPSG 的基础
ARP 欺骗	冒充别的计算机或者网关进行 ARP 应答，实现中间人攻击	在 DHCP Snooping 基础上，启用 DAI
IP 地址欺骗	冒充别的计算机 IP 地址发送 DOS 攻击数据包	在 DHCP Snooping 基础上，启用 IPSG
VLAN 跳跃攻击	把数据帧的 VLAN 标签封装为另一个 VLAN，导致跨 VLAN 的访问	禁止端口的 Trunk 协商，把 Native VLAN 设为不使用的 VLAN
同一 VLAN 中的设备之间的攻击	同一 VLAN 中的计算机是可以通信的，导致一旦一台主机被攻陷，其他计算机也受到威胁	启用端口保护，或者使用 PVLAN，详见 2.9
STP 根攻击	通过发送更高优先级的 BPDU，成为 STP 根桥，改变 STP 树拓扑	启用根保护，详见 3.6
VTP 攻击	发送伪造的 VTP 信息，覆盖正常的 VLAN 信息	VTP 配置密码，详见 2.7

6.1.1 交换机的访问安全

为了防止交换机被攻击者探测或者控制,必须对交换机进行基本的安全配置,具体有:
① 配置合格的密码,包括 Console 和 vty 密码,设置防止猜测密码措施;
② 使用 ACL,限制管理访问;
③ 配置系统警告用语;
④ 禁用不需要的服务;
⑤ 关闭 CDP;
⑥ 启用系统日志;
⑦ 使用 SSH 替代 Telnet;
⑧ 关闭 SNMP 或者使用 SNMP V3。

6.1.2 交换机的端口安全

交换机依赖 MAC 地址表转发数据帧,如果 MAC 条目不存在,则交换机将帧转发到交换机上的每一个端口(泛洪)。然而 MAC 地址表的大小是有限的,MAC 泛洪攻击利用这一限制用虚假源 MAC 地址轰炸交换机,直到把交换机 MAC 地址表变满。交换机随后进入称为"失效开放"(Fail-open)的模式,开始像集线器一样工作,将数据包泛洪到网络中的所有机器。因此,攻击者可看到发送到无 MAC 地址表条目的另一台主机的所有帧。要防止 MAC 泛洪攻击,可以配置端口安全特性,限制端口上所允许的有效 MAC 条目的数量,并定义当攻击发生时端口的动作:关闭、保护、限制。

6.1.3 DHCP Snooping——防 DHCP 欺骗

在局域网内,经常使用 DHCP 服务器为用户分配 IP 地址,DHCP 服务是一个没有认证的服务,即客户端和服务器端无法互相进行合法性的验证。在 DHCP 工作原理中,客户端以广播的方法来寻找服务器,并且只采用第一个达到的网络配置参数。如果在网络中存在多台 DHCP 服务器(有一台或更多台是非授权的),谁先应答,客户端就采用其提供的网络配置参数。假如非授权的 DHCP 服务器先应答,这样客户端获得的网络参数即是非授权的,客户端可能获取不正确的 IP 地址、网关和 DNS 等信息。在实际攻击中,攻击者还很可能恶意从授权的 DHCP 服务器上反复申请 IP 地址,导致授权的 DHCP 服务器消耗了全部地址,出现 DHCP 饥饿。

DHCP Snooping 可以防止 DHCP 饥饿和 DHCP 欺骗攻击。DHCP Snooping 截获交换机端口的 DHCP 应答报文,建立一张包含用户 MAC 地址、IP 地址、租用期、VLAN ID 和交换机端口等信息的表,并且 DHCP Snooping 还将交换机的端口分为可信任端口和不可信任端口,当交换机从一个不可信任端口收到 DHCP 服务器的报文时,比如 DHCP OFFER 报文、DHCP ACK 报文和 DHCP NAK 报文,交换机会直接将该报文丢弃;对从信任端口收到的 DHCP 服务器的报文,交换机不会丢弃而直接转发。一般将与用户相连的端口定义为不可信任端口,而将与 DHCP 服务器或者其他交换机相连的端口定义为可信任端口,也就是说,当在一个不可信任端口连接有 DHCP 服务器时,该服务器发出的报文将不能通过交换机的端口。因此只要将用户端口设置为不可信任端口,就可以有效地防止非授权用户私自设置 DHCP 服务而引起的 DHCP 欺骗。

6.1.4　DAI——防 ARP 欺骗

ARP 协议是用来获取目的计算机或者网关 MAC 地址的,通信发起方以广播方式发送 ARP 请求,拥有目的 IP 地址或者网关 IP 地址的工作站给予 ARP 应答,回送自己的 IP 地址和 MAC 地址。ARP 协议支持一种无请求 ARP 功能,同一网段上的所有工作站收到主动 ARP 广播后会将发送者的 MAC 地址和其宣布的 IP 地址保存,覆盖以前 Cache 的同一 IP 地址和对应的 MAC 地址。由于 ARP 无任何身份真实校验机制,攻击者发送误导的主动式 ARP 使网络流量经过恶意攻击者的计算机,攻击者就成为了通信双方的中间人,达到窃取甚至篡改数据的目的。攻击者发送的主动式 ARP 采用发送方私有 MAC 地址而非广播地址,通信接收方根本不会知道自己的 IP 地址被取代。

动态 ARP 检查(Dynamic ARP Inspection,DAI)可以防止 ARP 欺骗,它可以帮助保证接入交换机只传递"合法的"ARP 请求和应答信息。DAI 基于 DHCP Snooping 来工作,DHCP Snooping 监听绑定表,包括 IP 地址与 MAC 地址的绑定信息,并将其与特定的交换机端口相关联,动态 ARP 检测(DAI-Dynamic ARP Inspection)可以用来检查所有非信任端口的 ARP 请求和应答(主动式 ARP 和非主动式 ARP),确保应答来自真正的 MAC 地址所有者。交换机通过检查端口纪录的 DHCP 绑定信息和 ARP 应答的 IP 地址决定其是否是真正的 MAC 地址所有者,不合法的 ARP 包将被拒绝转发。

DAI 针对 VLAN 配置,对于同一 VLAN 内的端口,可以开启 DAI 也可以关闭 DAI,如果 ARP 包是从一个可信任的端口接收到的,就不需要做任何检查;如果 ARP 包是从一个不可信任的端口上接收到的,该包就只能在绑定信息被证明合法的情况下才会被转发出去。这样,DHCP Snooping 对于 DAI 来说也成为必不可少的。DAI 是动态使用的,相连的客户端主机不需要进行任何设置上的改变。对于没有使用 DHCP 的服务器,个别机器可以采用静态添加 DHCP 绑定表或 ARP access-list 的方法实现。

另外,通过 DAI 可以控制某个端口的 ARP 请求报文频率。一旦 ARP 请求频率超过预先设定的阈值,立即关闭该端口。该功能可以阻止网络扫描工具的使用,同时对有大量 ARP 报文特征的病毒或攻击也可以起到阻断作用。

6.1.5　IPSG——防 IP 地址欺骗

IP 地址欺骗是黑客进行 DOS 攻击时经常同时使用的一种手段。IPSG(IP Source Guard,IP 源防护)是一种基于 IP/MAC 的端口流量过滤技术,它可以防止局域网内的 IP 地址欺骗攻击。IPSG 能够确保二层网络中终端设备的 IP 地址不会被劫持,而且还能确保非授权设备不能通过自己指定 IP 地址的方式来访问网络或攻击网络导致网络崩溃及瘫痪。

交换机内部有一个 IP 源绑定表(IP Source Binding Table)作为每个端口接收到的数据包

的检测标准，只有在两种情况下，交换机会转发数据：所接收到的 IP 包满足 IP 源绑定表中 Port/IP/MAC 的对应关系；所接收到的是 DHCP 数据包，其余数据包将被交换机丢弃。

IPSG 也是基于 DHCP Snooping 进行工作的，交换机从 DHCP 监听绑定表（DHCP Snooping Binding Table）自动学习获得端口上绑定的 MAC 地址和 IP 地址。连接在交换机上的所有 PC 都被配置为动态获取 IP 地址，PC 作为 DHCP 客户端通过广播发送 DHCP 请求，DHCP 服务器将含有 IP 地址信息的 DHCP 回复通过单播的方式发送给 DHCP 客户端，交换机从 DHCP 报文中提取关键信息（包括 IP 地址、MAC 地址、VLAN 号、端口号和租期等）并把这些信息保存到 DHCP 监听绑定表中。

交换机根据 DHCP 监听绑定表的内容自动生成 IP 源绑定表，然后交换机根据 IP 源绑定表里面的内容自动在端口加载基于端口的 ACL，由该 ACL 过滤所有 IP 流量。在客户端发送的 IP 数据包中，只有其源 IP 地址和 MAC 地址满足源 IP 绑定表才会被发送，对于具有源 IP 绑定表之外的其他源 IP 地址的流量，都将被过滤。

6.1.6 VLAN 跳跃攻击

VLAN 跳跃攻击主要有以下 2 种方式。

1. IEEE 802.1q 和 ISL 标记攻击

IEEE 802.1q 和 ISL 标记攻击主要是利用管理员没有明确在端口上配置 **"switch mode access"** 的疏漏。交换机端口默认时可能为 DTP（Dynamic Trunk Protocol）auto 或者 DTP desirable，攻击者发送 DTP 协商包，那么端口将成为 Trunk 端口，就能接收通往任何 VLAN 的流量。由此，攻击者可以通过所控制的端口与其他 VLAN 通信。

对于这种攻击，只需将所有不可信的端口模式设置为 access 模式，即可预防这种攻击的侵袭。

2. 双标签

攻击者将带有两个标签的帧通过 Trunk 链路发送到另一个交换机上，对方交换机剥离一个标签后，数据帧里还有一个标签。交换机会把数据包转发到那个标签所指明的 VLAN，攻击者实现了从一个 VLAN 访问另一个 VLAN 的目的。

对于这种攻击，可以把 Trunk 链路上的 Native VLAN 设置为一个不存在的 VLAN，并禁止这个 VLAN 的数据从 Trunk 链路通过。

6.1.7 AAA

AAA 是 Authentication、Authorization、Accounting（认证、授权、审计）的简称，是网络安全的一种管理机制，提供了认证、授权、审计 3 种安全功能。

AAA 一般采用客户机/服务器结构，客户端运行于 NAS（Network Access Server，网络接入服务器）上，服务器上则集中管理用户信息。NAS 对于用户来讲是服务器端，对于 AAA 服务器来说是客户端。AAA 基本组网结构示意图如图 6-1 所示。

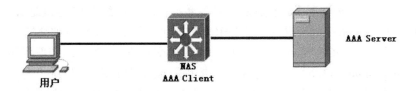

图 6-1　AAA 基本组网结构示意图

当用户想要通过某网络与 NAS（通常是网络设备）建立连接，从而获得访问其他网络的权利或取得某些网络资源的权利时，NAS 起到了验证用户的作用。NAS 负责把用户的认证、授权和审计信息传递给 AAA 服务器（RADIUS 服务器或 TACACS+ 服务器），RADIUS 协议或 TACACS+ 协议规定了 NAS 与服务器之间如何传递用户信息。

这 3 种安全服务功能的具体作用如下所述。

① 认证：确认远端访问用户的身份，判断访问者是否为合法的网络用户，即你是谁？

② 授权：对不同用户赋予不同的权限，限制用户可以使用的服务。例如，当用户成功登录交换机后，管理员可以授权用户对交换机进行配置，即你可以做什么？

③ 审计：记录用户使用网络服务中的所有操作，包括使用的服务类型、起始时间和数据流量等，它不仅是一种安全手段，也可以对用户实现计费，即你做了什么？

当然，用户可以只使用 AAA 提供的一种或多种安全服务。例如，公司仅仅想让员工在访问某些特定资源的时候进行身份认证，那么网络管理员只要在认证服务器配置就可以了。但是若希望对员工使用网络的情况进行记录，那么还需要配置审计服务器。

6.1.8　dot1x

虽然在二层交换机上可以采取端口安全、DHCP Snooping、DAI 和 IPSG 等安全措施，但还是不能解决用户非法接入网络的问题，一旦接入者把计算机的 MAC 地址改为合法的 MAC 地址，以上措施全部无效。dot1x 可以解决接入者的身份验证问题，dot1x 利用 AAA 服务，只有合法的用户才能接入到交换机上。

IEEE 802.1x 认证过程如图 6-2 所示，具体描述如下：

① 客户端向接入设备（交换机或者无线 AP）发送一个 EAPoL-Start 报文（认证请求），开始 IEEE 802.1x 认证接入。

② 接入设备向客户端发送 EAP-Request/Identity 报文（认证应答），要求客户端将用户名送上来。

③ 客户端回应一个 EAP-Response/Identity 报文给接入设备，其中包括用户名。

④ 接入设备将 EAP-Response/Identity 报文封装到 RADIUS Access-Request 报文中，发送给认证服务器（RADIUS 服务器）。

⑤ 认证服务器产生一个 Challenge，通过接入设备将 RADIUS Access-Challenge 报文发送给客户端，其中包含有 EAP-Request/MD5-Challenge。

⑥ 接入设备通过 EAP-Request/MD5-Challenge 发送给客户端，要求客户端进行认证。

⑦ 客户端收到 EAP-Request/MD5-Challenge 报文后，将密码和 Challenge 采用 MD5 算法处理后的 Challenged-Pass-word，在 EAP-Response/MD5-Challenge 中回应给接入设备。

⑧ 接入设备将 Challenge、Challenged Password 和用户名一起送到 RADIUS 服务器，由 RADIUS 服务器进行认证。

⑨ RADIUS 服务器根据用户信息，采用 MD5 算法判断用户是否合法，然后回应认证成功/失败报文到接入设备。如果成功，携带协商参数，以及用户的相关业务属性给用户授权；如果认证失败，则流程到此结束。

⑩ 如果认证通过，接入设备发起计费开始请求给 RADIUS 认证服务器。

⑪ RADIUS 认证服务器回应计费开始请求报文，用户接入到接入设备。

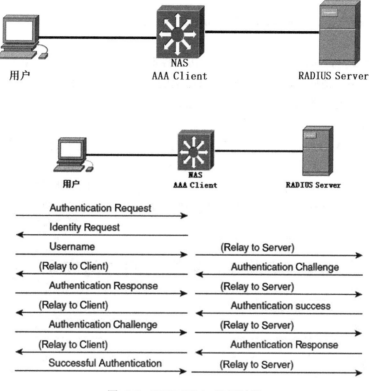

图 6-2 IEEE 802.1x 认证过程

6.1.9 SPAN

SPAN（Switched Port Analyzer）或者 RSPAN（Remote SPAN）经常用于监控网络流量，特别是用于入侵检测。SPAN 或 RSPAN 可以将一个交换机端口的流量镜像至另一个端口，即把端口的收发流量复制到该交换机或者其他交换机的另一个端口，这样，只需将分析或侦听设备连接至监控端口，即可实现对被监听端口的分析和侦听。由于 SPAN 采用复制（或镜像）源端口或源 VLAN 上的接收或发送（或两者都有的）流量到目的端口的方式，因此，SPAN 不影响源端口或 VLAN 的网络交换。除非特殊配置，除了 SPAN 或 RSPAN 会话的流量，目的端口不参与其他二层协议。目的端口只能是单独的一个实际物理端口，一个目的端口只能在一个 SPAN 会话中使用。

SPAN 拥有以下 3 种模式。

① 本地 SPAN：也称为 PSPAN，是指基于端口的 SPAN。源端口和目标端口都处于同一交换机，并且源端口可以是一个或多个交换机端口。

② 远程 SPAN 或者 RSPAN：目的端口和源端口不位于同一交换机上。这是一项高级功能，要求有专门的 VLAN 来传送该业务，并由交换机之间的 SPAN 进行监控，因此，要求中间交换机必须支持 RSPAN VLAN 技术。

③ VSPAN：指基于 VLAN 的 SPAN。SPAN 的源端口不是物理端口，而是 VLAN。在给定的交换机中，用户可以使用单个命令来选择对属于指定 VLAN 的所有端口进行监控。

6.1.10　RACL、VACL 和 MAC ACL

为确保网络安全，可以在交换机上应用访问控制列表。有以下 3 种 ACL。

① RACL：路由访问控制列表，与路由器上的 ACL 是一样的。可以应用在三层交换机上的三层端口（VLAN 端口、物理端口和 EtherChannel）上，也可以应用在二层端口上（但只能在 IN 方向）。

② VACL：VLAN 访问控制列表，应用在 VLAN 上，可以把流量与 RACL 和 MAC ACL 进行匹配，采取转发或者丢弃的动作。VACL 没有方向性，对进出 VLAN 的流量都生效。

③ MAC ACL：MAC 访问控制列表，可以应用在二层端口上。MAC ACL 可以根据帧的源 MAC 地址、目的 MAC 地址和帧的类型值转发或者过滤数据帧，而 MAC ACL 只能对非 IP 流量进行过滤。

6.2　实验 1：交换机的访问安全

1. 实验目的

通过本实验可以掌握：
① 交换机访问安全的配置。
② 用 SSH 协议访问交换机的方法。

2. 实验拓扑

交换机访问安全实验拓扑如图 6-3 所示。

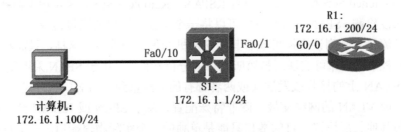

图 6-3　交换机访问安全实验拓扑

3. 实验步骤

（1）准备工作：配置 IP 地址

R1(config)#**interface GigabitEthernet0/0**
R1(config-if)#**no shutdown**
R1(config-if)#**ip address 172.16.1.200 255.255.255.0**

S1(config)#**interface vlan 1**
S1(config-if)#**no shutdown**
S1(config-if)#**ip address 172.16.1.1 255.255.255.0**

将计算机的网卡配置为 172.16.1.100/255.255.255.0，测试计算机和交换机 S1、路由器 R1 和交换机 S1 的连通性。

（2）在交换机 S1 上配置密码、login 安全等

S1(config)#**enable secret cisco**
//配置 enable 的密码，建议不要使用 "enable password" 命令，因为它的密码不加密存放
S1(config)#**service password-encryption**
//启用密码加密服务，这样以后配置的密码将加密存放，然而需要强调的是：和 "enable secret" 的密码不一样的是，采用这种方式的加密是可逆的，思科网站提供破解程序
S1(config)#**line console 0**
S1(config-line)#**password cisco**
S1(config-line)#**login**   //配置 Console 口的密码
S1(config)#**line vty 0 15**
S1(config-line)#**password cisco**
S1(config-line)#**login**

以上命令配置 Telnet 或者 SSH 的密码和登录方式。采用上面的配置方法没有办法区分不同的用户。如果需要有不同用户，他们有各自的密码，可以采用 AAA 服务器，参见本章后相关内容。

> **提示**
> 以上为了说明方便，使用 "cisco" 作为密码，在实际应用中，密码应该满足复杂性要求，即密码中至少包含大写字母、小写字母、数字和特殊字符中的三样，长度不得少于 8 位。

S1(config)#**line console 0**
S1(config-line)#**exec-timeout 5 30**
//配置超时时间，当用户在 5 分 30 秒内没有任何输入时，将被自动注销，这样可以减少不安全因素
S1(config)#**service tcp-keepalives-in**
//使得交换机当没有收到远程系统的响应时会自动关闭连接，减少受到 DOS 攻击的机会
S1(config)#**access-list 5 permit 172.16.0.0 0.0.255.255**
//定义了访问控制列表，允许 172.16.0.0/16 网段的计算机远程访问
S1(config)#**line vty 0 15**
S1(config-line)#**access-class 5 in**
//限定 ACL 5 指定的计算机 Telnet（或者 SSH）该交换机，这样大大减少被攻击的机会
S1(config-line)#**exec-timeout 5 30**
S1(config-line)#**login local**

//配置 Telnet 或者 SSH 登录进行认证，则当用户从 Telnet 或者 SSH 登录时会要求输入用户名和密码，如下所示

User Access Verification
Username:

为了防止黑客暴力破解密码，我们可以对用户频繁尝试登录进行限制，如下所示：

S1(config)#**login block-for 60 attempts 3 within 30**
//配置用户 30 秒内连续登录失败 3 次后，等待 60 秒后（进入安静期）才能再次登录。这样配置后，黑客在 30 秒内只能猜测 2 个密码，失去暴力破解的前提

S1(config)#**login delay 10**   //配置用户登录成功后，10 秒后才能再次登录

S1(config)#**access-list 10 permit host 172.16.1.150**

S1(config)#**login quiet-mode access-class 10**

在之前的配置中，当用户多次输错密码后，交换机将进入 60 秒的安静期，禁止登录。但是以上配置设置了例外的情况，安静期内 ACL 10 指定的计算机仍然可以登录，通常指定的计算机是管理员的计算机。这样的目的是防止黑客损人不利己的做法：我登录不了，也不让其他人登录。

S1(config)#**login on-failure log**   //配置登录失败会在日志中记录

S1(config)#**login on-success log**   //配置登录成功会在日志中记录

S1#**show login**

    A login delay of 10 seconds is applied.

    Quiet-Mode access list 10 is applied.

    All successful login is logged.

    All failed login is logged.

    Router enabled to watch for login Attacks.

    If more than 3 login failures occur in 30 seconds or less,

    logins will be disabled for 60 seconds.

    Router presently in Normal-Mode.

    Current Watch Window remaining time 7 seconds.

    Present login failure count 0.

//以上显示和 login 有关的全部配置情况

我们常常配置一些提示信息，在用户登录时显示给管理员，如下所示：

S1(config)#**banner motd #**

Without authorization, shall not be visited

#

以上配置了每日信息，除此之外还有其他一些信息，如下所述。

① **banner motd**：在交换机提示输入用户和密码前出现。

② **banner login**：在提示输入用户和密码前出现，但是是在 banner motd 之后。

③ **banner exec**：在进入 exec 模式之前提示的信息，即用户通过了认证后。

继续配置如下：

S1(config)#**banner exec #**

```
=========
Welocome!
=========
#
```
S1(config)#**banner login #**
```
-----------------
Switch1.szpt.edu.cn
-----------------
#
```

用户登录时提示的信息如下所示：

```
-----------------------------------
Without authorization, shall not be visited
-----------------------------------

-----------------
Switch1.szpt.edu.cn
-----------------
User Access Verification
Password:
=========
Welocome!
=========
S1>
```

（3）配置 SSH

通过 Telnet 访问交换机是一种简单方法，但是这是一种很不安全的方法，因为 Telnet 协议是采用明文传输的。我们可以使用 Telnet 的替代者 SSH 协议，该协议加密传输数据，端口号为 22。配置如下所示：

S1(config)#**clock timezone CST 8**
//配置交换机的时区，中国时间简写为 CST，为东八时区，比标准时间早 8 小时
S1#**clock set 8:32:00 5 FEB 2018**   //配置交换机的时间，注意该命令在特权模式下直接配置
S1(config)#**ip domain-name cisco.com**
//配置域名，如果你单位有自己的域名，请使用它，否则随便配置一个
S1(config)#**crypto key generate rsa**   //产生加密密钥
% You already have RSA keys defined named S1.cisco.com.
% Do you really want to replace them? [yes/no]: yes
//提示是否覆盖原有的密钥
Choose the size of the key modulus in the range of 360 to 2048 for your
    General Purpose Keys. Choosing a key modulus greater than 512 may take
    a few minutes.

How many bits in the modulus [512]:1024   //密钥的长度
000056: Feb 5 00:33:01: %SSH-5-ENABLED: SSH 1.99 has been enabled
% Generating 1024 bit RSA keys, keys will be non-exportable...

[OK] (elapsed time was 4 seconds)

S1(config)#**username user1 secret cisco**　　　//SSH 需要用户名和密码，创建用户
S1(config)#**line vty 0 15**
S1(config)# **ip ssh version 2**　　　　　　　　//使用 SSH 的版本 2
S1(config-line)#**transport input ssh**
//只允许用户通过 SSH 远程登录到交换机。默认是 transport input all，因此默认时可以使用 Telnet 协议
S1(config-line)#**login local**　　　　　　　　//从交换机上本地验证用户名和密码

从路由器 R1 远程登录到 S1 上，如下所示：

R1#**ssh -v 2 -c aes128-cbc -l user1 172.16.1.1**
Password:

Without authorization, shall not be visited

//这里 R1 是 SSH 的客户端。SSH 命令格式如下所示：
R2#ssh ?
　　-c　　Select encryption algorithm　　//指明加密算法
　　-l　　Log in using this user name　　//指明用户名
　　-m　　Select HMAC algorithm　　//指明 Hash 算法
　　-o　　Specify options　　//指明其他选项
　　-p　　Connect to this port　　//指明服务器上的 SSH 端口，默认值为 22
　　-v　　Specify SSH Protocol Version　　//指明 SSH 版本，默认为 Version1
　　WORD　　IP address or hostname of a remote system　　//服务器 IP 地址或者主机名

当然也可以从计算机上用 SSH 远程登录到交换机，本书第 1 章介绍的 SecureCRT 就支持 SSH。从"文件"→"快速连接"菜单打开图 6-4 所示窗口，在"协议"处选择 SSH2，在"主机名"处输入 IP 地址，在"用户名"处输入用户名，其他选项可自行选择或者保持默认，单击"连接"按钮，打开图 6-5 所示窗口，单击"接受并保存"接收和保存服务器发送来的 key。在图 6-6 所示窗口中输入用户名和密码即可。如图 6-7 所示，用 SSH 成功连接交换机。

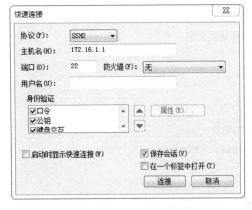

图 6-4　快速连接窗口

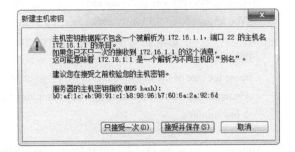

图 6-5　保存服务器发送来的 key

第 6 章 交换机的安全

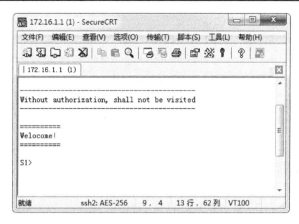

图 6-6 输入用户名和密码　　　　图 6-7 用 SSH 成功连接交换机

（4）关闭不必要的服务

默认时，交换机有各种各样的服务（使用 TCP 或者 UDP 端口）已经被开启。在不影响使用的情况下，为了安全起见可以把不必要的服务关闭，操作如下：

```
S1(config)#no cdp run       //关闭 CDP，CDP 协议是 Cisco 的邻居发现协议
S1(config)#no lldp run      //关闭 LLDP，LLDP 协议可以说是 CDP 的国际标准版
S1(config)#no ip source-route
//关闭基于源的路由功能。该功能指用户可以在发送出的 IP 数据包中指明转发路径
S1(config)#no ip http server
//关闭 HTTP 服务功能，关闭后我们就无法用浏览器配置交换机了
S1(config)#no service tcp-small-servers
//关闭 TCP 端口号小于或者等于 19 的服务，例如，datetime、echo 和 chargen 等服务
S1(config)#no service udp-small-servers
//关闭 UDP 端口号小于或者等于 19 的服务
S1(config)#no service finger
//关闭 finger 服务，该服务和"show user"命令很类似，用于显示当前登录到交换机的用户
S1(config)#no service dhcp
//关闭 DHCP 服务，DHCP 用于为计算机分配 IP 地址等。如果确实需要，请保留
S1(config)#no ip name-server
S1(config)#no ip domain-lookup
//关闭交换机作为 DNS 客户端的功能，即交换机不再到 DNS 服务器上去查询主机的 IP 地址了
S1(config)#no service config
//关闭交换机在网络上查找配置文件的功能。默认时，交换机开机后如果在 NVRAM 找不到配置文件，
将广播查找 TFTP 服务器
S1(config)#no snmp-server
//删除所有和 SNMP 配置有关的配置。SNMP 称为简单网络管理协议，主要配置和监控网络设置。如果
确实需要，请保留
```

（5）配置 HTTPS

在步骤（4）中，关闭了 HTTP 服务，如果需要使用浏览器配置交换机，可以配置 HTTPS

服务，如下所示：

```
S1(config)#ip http secure-server
% Generating 1024 bit RSA keys, keys will be non-exportable...
[OK] (elapsed time was 3 seconds)
*Mar   1 20:14:08.288: %SSH-5-ENABLED: SSH 1.99 has been enabled
*Mar   1 20:14:08.548: %PKI-4-NOAUTOSAVE: Configuration was modified.  Issue "write memory" to save new certificate
//HTTPS 需要 1 024 位的证书保护通信，启用 HTTPS 自动产生证书，记得要执行"write"或者"copy running-config startup-config"命令保存配置以及产生的证书
S1(config)#ip http authentication local
//HTTPS 服务需要用户名和密码，这里采用本地认证。也可以采用 AAA，见本章后面相关内容
S1(config)#username user2 privilege 15 secret cisco
//创建一个用户，用户要有等级为 15 的权限才能配置交换机
```

6.3 实验 2：交换机端口安全

1. 实验目的

通过本实验可以掌握：

① 如何管理交换机的 MAC 地址表。
② 交换机端口安全功能的配置。

2. 实验拓扑

交换机端口安全实验拓扑如图 6-8 所示。

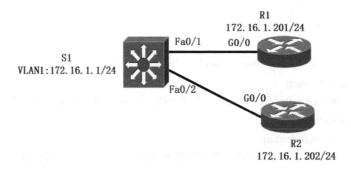

图 6-8 交换机端口安全实验拓扑

3. 实验步骤

交换机端口安全特性可以让我们配置交换机端口，使得当具有非法 MAC 地址的设备接入时，交换机自动关闭端口或者拒绝非法设备接入，也可以限制某个端口上最大的 MAC 地址数量。这里我们限制交换机 S1 的 Fa0/1 端口，只允许 R1 接入。

（1）配置交换机上的 MAC 地址表

```
R1(config)#interface GigabitEthernet0/0
```

R1(config-if)#**no shutdown**
R1(config-if)#**ip address 172.16.1.201 255.255.255.0**

R2(config)#**interface GigabitEthernet0/0**
R2(config-if)#**no shutdown**
R2(config-if)#**ip address 172.16.1.202 255.255.255.0**

S1(config)#**interface vlan 1**
S1(config-if)#**ip address 172.16.1.1 255.255.255.0**

R1#**show interfaces GigabitEthernet0/0**
GigabitEthernet0/0 is up, line protocol is up
   Hardware is CN Gigabit Ethernet, address is f872.ead6.f4c8 (bia f872.ead6.f4c8)
//这里可以看到 R1 的 G0/0 端口的 MAC 地址，记下它
 Internet address is 172.16.1.201/24
   MTU 1500 bytes, BW 100000 Kbit/sec, DLY 100 usec,
（此处省略部分输出）

R2#**show interfaces GigabitEthernet0/0**
GigabitEthernet0/0 is up, line protocol is up
   Hardware is CN Gigabit Ethernet, address is f872.ea69.1c78 (bia f872.ea69.1c78)
//这里可以看到 R2 的 G0/0 端口的 MAC 地址，记下它

S1#**show mac address-table**
       Mac Address Table

VLAN	Mac Address	Type	Ports	
----	-----------	--------	-----	
（此处省略部分输出）				
1	d0c7.89c2.3119	DYNAMIC	Fa0/23	
1	d0c7.89c2.311a	DYNAMIC	Fa0/24	
1	d0c7.89c2.3140	DYNAMIC	Fa0/23	
1	d0c7.89c2.8383	DYNAMIC	Fa0/22	
1	f872.ea69.18b8	DYNAMIC	Fa0/3	
1	**f872.ea69.1c78**	DYNAMIC	**Fa0/2**	//R2 路由器的 MAC 地址
1	**f872.ead6.f4c8**	DYNAMIC	**Fa0/1**	//R1 路由器的 MAC 地址

以上显示了交换机上的 MAC 地址表。VLAN 列：计算机所在的 VLAN；Mac Address 列：计算机的 MAC 地址；Type 列："DYNAMIC" 表示 MAC 记录是交换机动态学习到的，"STATIC" 表示 MAC 记录是静态配置或系统保留的；Ports 列：计算机所连接的交换机端口。

S1#**show mac address-table aging-time**
Global Aging Time: 300
Vlan   Aging Time

//查看 MAC 地址表的超时时间，默认值为 300 秒

S1(config)#**mac address-table aging-time 120 vlan 1**
//改变 VLAN1 上的 MAC 地址表的超时时间为 120 秒。
S1(config)#**mac address-table static f872.ea69.1c78 vlan 1 interface FastEthernet0/2**
//把路由器 R2 的 MAC 地址静态添加到 MAC 地址表中
S1#**show mac address-table**

            Mac Address Table

Vlan   Mac Address   Type   Ports
----   -----------   --------   -----
(此处省略部分输出)
  1   d0c7.89c2.3119   DYNAMIC   Fa0/23
  1   d0c7.89c2.311a   DYNAMIC   Fa0/24
  1   d0c7.89c2.3140   DYNAMIC   Fa0/23
  1   d0c7.89c2.8383   DYNAMIC   Fa0/22
  1   f872.ea69.18b8   DYNAMIC   Fa0/3
  1   f872.ea69.1c78   STATIC   Fa0/2   //这是静态记录
  1   f872.ead6.f4c8   DYNAMIC   Fa0/1

//静态配置了 MAC 地址表，该记录的类型为 STATIC，永不从 MAC 表中超时（端口关闭了，该记录还是会被删除的）。对于服务器等位置较为固定的计算机，为了安全起见，建议配置静态 MAC 地址表
S1(config)#**no mac address-table static f872.ea69.1c78 vlan 1 interface FastEthernet0/2**
//删除静态配置的 MAC 地址表

（2）配置交换机端口安全（静态安全 MAC 地址）

S1(config)#**interface FastEthernet0/1**
S1(config-if)#**shutdown**
S1(config-if)#**switchport mode access**
//把端口改为访问模式，即用来接入计算机
S1(config-if)#**switchport port-securitiy**   //打开交换机的端口安全功能
S1(config-if)#**switchport port-securitiy maximum 1**
//只允许该端口下的 MAC 条目最大数量为 1，即只允许一个设备接入，实际上，这是默认值
S1(config-if)#**switchport port-securitiy violation shutdown**
//配置攻击发生时端口要采取的动作：关闭端口。实际上，这是默认值

 提示

switchport port-securitiy violation { protect | shutdown | restrict }命令描述如下。
① **protect**：当新的计算机接入时，如果该端口的 MAC 条目超过最大数量，则这个新的计算机将无法接入，而原有的计算机不受影响，交换机也不发送警告信息。
② **restrict**：当新的计算机接入时，如果该端口的 MAC 条目超过最大数量，则这个新的计算机无法接入，而原有的计算机不受影响，交换机将发送警告信息（在日志中可以看到），也会对违规行为计数。

③ **shutdown**：当新的计算机接入时，如果该端口的 MAC 条目超过最大数量，则该端口将会被关闭，则这个新的计算机和原有的计算机都无法接入，需要管理员使用 "**shutdown**" 和 "**no shutdown**" 命令重新打开。表 6-2 是交换机端口安全违规模式比较。

表 6-2 交换机端口安全违规模式比较

违规模式	转发违规之前计算机的流量	转发造成违规计算机的流量	发出警告	显示错误消息	增加违规计数	关闭端口
Protect	是	否	否	否	否	否
Restrict	是	否	是	否	是	否
Shutdown	否	否	是	否	是	是

```
S1(config-if)#switchport port-security mac-address f872.ead6.f4c8
//允许路由器 R1 从 GigabitEthernet0/1 端口接入
S1(config-if)#no shutdown
S1#show mac-address-table
          Mac Address Table
-------------------------------------------
Vlan    Mac Address       Type        Ports
----    -----------       --------    -----
(此处省略部分输出)
 1      d0c7.89c2.3119    DYNAMIC     Fa0/23
 1      d0c7.89c2.311a    DYNAMIC     Fa0/24
 1      d0c7.89c2.3140    DYNAMIC     Fa0/23
 1      d0c7.89c2.8383    DYNAMIC     Fa0/22
 1      f872.ea69.18b8    DYNAMIC     Fa0/3
 1      f872.ea69.1c78    DYNAMIC     Fa0/2
 1      f872.ead6.f4c8    STATIC      Fa0/1      //路由器 R1 的 MAC 地址
Total Mac Addresses for this criterion: 27
//R1 的 MAC 地址已经被登记在 Fa0/1 端口，并且表明是静态加入的
```

这时从 R1 上 ping 交换机的管理地址，可以 ping 通，如下所示：

```
R1#ping 172.16.1.1
 Type escape sequence to abort.
Sending 5, 100-byte ICMP Echos to 172.16.1.1, timeout is 2 seconds:
!!!!!
Success rate is 100 percent (5/5), round-trip min/avg/max = 1/1/4 ms
```

（3）模拟非法接入

我们将 R1 端口的 MAC 地址改成另一个 MAC 地址，这样就可以模拟从交换机的 Fa0/1 端口接入另一计算机了。如下所示：

```
R1(config)#interface GigabitEthernet0/0
R1(config-if)#shutdown
R1(config-if)#mac-address 12.12.12
R1(config-if)#no shutdown
```

几秒钟后，则在 S1 上出现：

　　*Mar　1 00:23:55.240: %LINK-3-UPDOWN: Interface FastEthernet0/1, changed state to up
　　*Mar　1 00:23:56.247: %LINEPROTO-5-UPDOWN: Line protocol on Interface FastEthernet0/1, changed state to up
　　*Mar　1 00:24:08.687: %PM-4-ERR_DISABLE: psecure-violation error detected on Fa0/1, putting Fa0/1 in err-disable state
　　*Mar　1 00:24:08.687: %PORT_SECURITY-2-PSECURE_VIOLATION: Security violation occurred, caused by MAC address 0012.0012.0012 on port FastEthernet0/1.
　　*Mar　1 00:24:09.694: %LINEPROTO-5-UPDOWN: Line protocol on Interface FastEthernet0/1, changed state to down
　　*Mar　1 00:24:10.692: %LINK-3-UPDOWN: Interface FastEthernet0/1, changed state to down
　　//以上提示 Fa0/1 端口被关闭。配置的最终效果是：交换机的 Fa0/1 端口只能让某一固定的设备（R1）接入

S1#show interface fastEthernet0/1

FastEthernet0/1 is down, line protocol is down (err-disabled)
　　Hardware is Fast Ethernet, address is d0c7.89ab.1183 (bia d0c7.89ab.1183)
　　MTU 1500 bytes, BW 10000 Kbit/sec, DLY 1000 usec,
　　　　reliability 255/255, txload 1/255, rxload 1/255
//以上表明 Fa0/1 端口因为错误而被关闭。非法设备移除后（在 R1 上的 G0/0 端口上，执行 "**no mac-address**"），在交换机的 Fa0/1 端口下执行 "**shutdown**" 和 "**no shutdown**" 命令可以重新打开该端口

下面试试另一违规模式 **restrict** 的效果：

　　R1(config)#**interface GigabitEthernet0/0**
　　R1(config-if)#**shutdown**
　　R1(config-if)#**no mac-address 12.12.12**
　　R1(config-if)#**no shutdown**
　　//先把 R1 的 G0/0 端口的 MAC 地址恢复为原来的地址
　　S1(config)#**interface fastEthernet0/1**
　　S1(config-if)#**switchport port-security violation restrict**
　　//把违规模式改为 restrict。测试从 R1 上 ping 172.16.1.1，应该正常 ping 通

　　R1(config)#**interface GigabitEthernet0/0**
　　R1(config-if)#**shutdown**
　　R1(config-if)#**mac-address 12.12.12**
　　R1(config-if)#**no shutdown**
　　//重新修改 R1 的 Fa0/0 端口的 MAC 地址，再测试从 R1 上 ping 172.16.1.1，不能 ping 通。同时交换机上会出现以下信息
　　*Mar　1 00:30:55.090: %PORT_SECURITY-2-PSECURE_VIOLATION: Security violation occurred, caused by MAC address 0012.0012.0012 on port FastEthernet0/1.
　　*Mar　1 00:31:01.364: %PORT_SECURITY-2-PSECURE_VIOLATION: Security violation occurred, caused by MAC address 0012.0012.0012 on port FastEthernet0/1.

结果表明：**restrict** 违规模式不影响原有的计算机通信，但是导致违规发生的计算机的通信不被允许。

第 6 章 交换机的安全

（4）配置交换机端口安全（动态安全 MAC 地址）

动态安全 MAC 地址和静态安全 MAC 地址的差别在于：前者少配置"**switchport port-security mac-address f872.ead6.f4c8**"命令。交换机完整的配置命令如下所示：

S1(config)#**default interface fastEthernet0/1**
//把 Fa0/1 端口上的配置恢复到默认配置（即出厂时的配置）
S1(config)#**interface fastEthernet0/1**
S1(config-if)#**shutdown**
S1(config-if)#**switchport mode access**
S1(config-if)#**switchport port-security**
S1(config-if)#**switchport port-security maximum 1**
S1(config-if)#**switchport port-security violation shutdown**
S1(config-if)#**no shutdown**

这样，当交换机 Fa0/1 端口接入第一台计算机时，该计算机的 MAC 地址将作为 STATIC 类型被添加到 MAC 地址表中。当接入第二台计算机时，由于交换机 Fa0/1 端口最大 MAC 地址数量为 1，该计算机将被认为是入侵者，交换机关闭该端口。最终的效果是：交换机的 Fa0/1 端口只能有一台计算机接入（但是并不限制是哪个 MAC 地址）。

测试步骤与步骤（3）类似。

（5）配置交换机端口安全（粘滞安全 MAC 地址）

静态安全 MAC 地址可以使得交换机的 Fa0/1 端口只能接入某一固定的计算机，然而需要使用"**switchport port-security mac-address f872.ead6.f4c8**"命令，这样就需要一一查出计算机的 MAC 地址，这是一项工作量巨大的事情，粘滞安全 MAC 地址可以解决这个问题。交换机完整的配置命令如下：

S1(config)#**default interface fastEthernet0/1**
S1(config)#**interface fastEthernet0/1**
S1(config-if)#**shutdown**

S1(config-if)#**switchport mode access**
S1(config-if)#**switchport port-security**
S1(config-if)#**switchport port-security maximum 1**
S1(config-if)#**switchport port-security violation shutdown**
S1(config-if)#**switchport port-security mac-address sticky**
S1(config-if)#**no shutdown**
//配置交换机端口自动粘滞 MAC 地址

从 R1 上 ping 交换机 172.16.1.1，然后在交换上进行如下操作

S1#**show running-config interface fastEthernet0/1**
Building configuration...
Current configuration : 188 bytes
!
interface FastEthernet0/1
 switchport mode access
 switchport port-security

switchport port-security mac-address sticky

  switchport port-security mac-address sticky f872.ead6.f4c8

//可以发现，交换机自动把 R1 的 MAC 地址粘滞在该端口下了，这时相当于执行了 "**switchport port-security mac-address f872.ead6.f4c8**" 命令。以后该端口只能接入路由器 R1

 提示

  实际工作中可以这样进行如下操作：
  S1(config)#**interface range fastEthernet0/1 -24**
  //批量配置 Fa0/1～Fa0/24 端口。在有的 IOS 中，破折号前需要有空格才行
  S1(config-if)#**shutdown**
  S1(config-if)#**switchport mode access**
  S1(config-if)#**switchport port-securitiy**
  S1(config-if)#**switchport port-securitiy maximum 1**
  S1(config-if)#**switchport port-securitiy violation shutdown**
  S1(config-if)#**switchport port-security mac-address sticky**
  S1(config-if)#**no shutdown**
  然后等交换机 Fa0/1 和 Fa0/24 端口上的计算机都开机后，在交换机上检查确认已经粘滞了 MAC 地址后，把配置保存下来（使用 "**copy running-config startup-config**" 命令）。

提示

  当交换机发现有入侵者而关闭了端口后，需要管理员手工重新打开端口。我们也可以通过配置使端口自动恢复：
  S1(config)#**errdisable recovery cause psecure-violation**
  //允许交换机自动恢复因为端口安全而关闭的端口
  S1(config)#**errdisable recovery interval 60**
  //配置交换机 60 秒后自动恢复端口

4. 实验调试

```
S1#show port-security
Secure Port  MaxSecureAddr   CurrentAddr    SecurityViolation   Security Action
              (Count)         (Count)           (Count)
---------------------------------------------------------------------------------
  Fa0/1          1               1                 0              Shutdown
---------------------------------------------------------------------------------
Total Addresses in System (excluding one mac per port)    : 0
Max Addresses limit in System (excluding one mac per port) : 6144
```

查看端口安全的设置情况，各列的含义已经很明了。

```
S1#show port-security interface fastethernet0/1
Port Security                : Enabled    //端口启用了安全模式
Port Status                  : Secure-up  //当前端口的状态
Violation Mode               : Shutdown   //违规模式
Aging Time                   : 0 mins
Aging Type                   : Absolute
SecureStatic Address Aging   : Disabled
```

Maximum MAC Addresses	: 1	//最大 MAC 地址数量
Total MAC Addresses	: 1	//当前端口的 MAC 地址数量
Configured MAC Addresses	: 0	//手工配置的 MAC 地址数量
Sticky MAC Addresses	: 1	//粘贴的 MAC 地址数量
Last Source Address:Vlan	: f872.ead6.f4c8:1	
Security Violation Count	: 0	//违规次数

6.4 实验 3：DHCP 欺骗

1. 实验目的

通过本实验可以掌握：
① DHCP 欺骗攻击的防范原理。
② 防范 DHCP 欺骗攻击的配置。

2. 实验拓扑

DHCP 欺骗实验拓扑如图 6-9 所示。R1 作为 DHCP 服务器，而 R2 和 R4 作为 DHCP Client，R3 模拟用户私自搭建的 DHCP 服务器。在交换机 S1 和 S2 上配置 DHCP 欺骗防范措施。

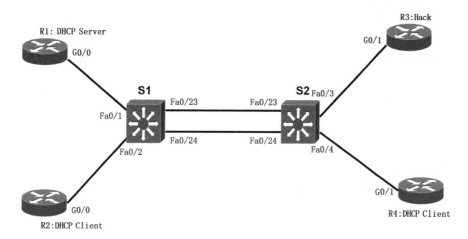

图 6-9 DHCP 欺骗实验拓扑

3. 实验步骤

（1）配置 R1 和 R3 作为 DHCP Server，配置 R2 和 R4 作为 DHCP Client

```
R1(config)#interface GigabitEthernet0/0
R1(config-if)#no shutdown
R1(config-if)#ip address 172.16.1.1 255.255.255.0
R1(config)#ip dhcp pool DHCP-TEST
R1(dhcp-config)#network 172.16.1.0 255.255.255.0
```

//配置 R1 为 DHCP 服务器

R3(config)#**interface GigabitEthernet0/1**
//R3 使用 G0/1 端口
R3(config-if)#**no shutdown**
R3(config-if)#**ip address 172.16.1.3 255.255.255.0**
R3(config)#**ip dhcp pool HACK**
R3(dhcp-config)#**network 172.16.1.0 255.255.255.0**
//配置 R3 为用户私自搭建的 DHCP 服务器

S1(config)#**interface range fastEthernet0/23 -24**
S1(config-if)#**switchport trunk encapsulation dot1q**
S1(config-if)#**switchport mode trunk**
S1(config)#**interface fastEthernet0/22**
S1(config-if)#**shutdown**

S2(config)#**interface range fastEthernet0/23 -24**
S2(config-if)#**switchport trunk encapsulation dot1q**
S2(config-if)#**switchport mode trunk**
S2(config)#**interface fastEthernet0/22**
S2(config-if)#**shutdown**
//配置 S1 和 S2 之间的链路为 Trunk 链路，同时关闭 Fa0/2（见第 1 章的图 1-1），避免影响实验

R2(config)#**interface GigabitEthernet0/0**
R2(config-if)#**no shutdown**
R2(config-if)#**ip address dhcp**

R4(config)#**interface GigabitEthernet0/1**
R4(config-if)#**no shutdown**
R4(config-if)#**ip address dhcp**
//配置 R2 的 G0/0 和 R4 的 G0/1 端口自动获得 IP 地址
R2#**show ip interface brief**
Interface	IP-Address	OK?	Method	Status	Protocol
Embedded-Service-Engine0/0	unassigned	YES	unset	administratively down	down
GigabitEthernet0/0	172.16.1.4	YES	DHCP	up	up
GigabitEthernet0/1	unassigned	YES	unset	administratively down	down

以上表明 R2 已经获得 IP 地址了，DHCP Server 已经正常工作。IP 地址之所以不是 172.16.1.1 是因为之前已经执行好几次 "**ip address dhcp**" 命令了。

 提示

请先关闭 R3 的 G0/1 端口，测试 R2 和 R4 已经正常从 R1 获取 IP 地址；再打开 R3 的 G0/1 端口并关闭 R1 的 G0/0 端口，测试 R2 和 R4 已经正常从 R3 获取 IP 地址。目的是确认 R1 和 R3 这两个 DHCP 服务器都能正常为 DHCP Client 分配 IP 地址。

第 6 章 交换机的安全

（2）在 S2 上配置防 DHCP 欺骗

交换机 S2 作为接入层交换机。

S2(config)#**ip dhcp snooping**　　//打开 S2 的 DHCP 监听功能
S2(config)#**ip dhcp snooping vlan 1**　　//配置 S2 监听 VLAN1 上的 DHCP 数据包
S2(config)#**no ip dhcp snooping information option**
//配置交换机 S2 不在 DHCP 报文中插入 option 82。option 82 是 DHCP 中继代理，稍后进一步讨论
S2(config)#**ip dhcp snooping verify mac-address**
//配置 S2 检查 DHCP 报文中的 MAC 地址和以太网帧的 MAC 地址是否相同
S2(config)#**interface fastEthernet0/23**
S2(config-if)#**ip dhcp snooping trust**
S2(config)#**interface fastEthernet0/24**
S2(config-if)#**ip dhcp snooping trust**
//交换机 S2 虽然并没有直接接在 DHCP 服务器上，但是这两个端口是上连端口，该端口会接收 DHCP 服务器的 DHCP 响应包，所以应该配置成信任端口

S2(config)#**ip dhcp snooping database flash:dhcp_snooping.db**
//将 DHCP 监听绑定表保存在 flash 中，文件名为 dhcp_snooping.db。之所以配置 database 是防止断电后记录丢失。如果记录很多，可以把 database 放到 TFTP 和 FTP 等服务器上
S2(config)#**ip dhcp snooping database write-delay 15**
//DHCP 监听绑定表发生更新后，等待 15 秒，再写入文件，默认值为 300 秒；可选范围为 15～86 400 秒
S2(config)#**ip dhcp snooping database timeout 15**
//DHCP 监听绑定表尝试写入操作失败后，重新尝试写入操作，直到 15 秒后停止尝试。默认值为 300 秒；可选范围为 0～86 400 秒

S2(config-if)#**interface range fastEthernet0/3 -4**
S2(config-if)#**ip dhcp snooping limit rate 10**
//默认时交换机的端口是非信任端口，因此如果端口是不可信任的就无须配置。这里限制端口每秒钟能接收的 DHCP 数据包数量为 10 个

（3）在 S1 上配置防 DHCP 欺骗

S1(config)#**ip dhcp snooping**
S1(config)#**ip dhcp snooping vlan 1**
S1(config)#**no ip dhcp snooping information option**
S1(config)#**ip dhcp snooping database flash:dhcp_snooping.db**
S1(config)#**ip dhcp snooping database write-delay 15**
S1(config)#**ip dhcp snooping database timeout 15**
S1(config)#**interface fastEthernet0/1**
S1(config-if)#**ip dhcp snooping trust**

S1 的 Fa0/1 端口接的是合法的 DHCP 服务器，该端口要配置成信任端口。其他端口是非信任端口。特别需要注意的是 Fa0/23 和 Fa0/24 端口是下连到交换机 S2 的接口，并不需要配置为信任端口。

（4）测试

R2(config)#**interface GigabitEthernet0/0**

R2(config-if)#**no ip address**
R2(config-if)#**ip address dhcp**
R2#**show ip interface brief**

请查看 R2 上的 IP 地址，应该能正常获得 IP 地址。

R4(config)#**interface GigabitEthernet0/1**
R4(config-if)#**no ip address**
R4(config-if)#**ip address dhcp**
R4#**show ip interface brief**

请查看 R4 上的 IP 地址，应该能正常获得 IP 地址。

S1#**show ip dhcp snooping binding**

MacAddress	IpAddress	Lease(sec)	Type	VLAN	Interface
F8:72:EA:69:1C:78	172.16.1.8	86303	dhcp-snooping	1	FastEthernet0/2
F8:72:EA:C8:4F:99	172.16.1.7	86378	dhcp-snooping	1	FastEthernet0/23

Total number of bindings: 2
//在交换机 S1 上看到了 DHCP 监听的绑定

S2#**show ip dhcp snooping binding**

MacAddress	IpAddress	Lease(sec)	Type	VLAN	Interface
F8:72:EA:C8:4F:99	172.16.1.7	86355	dhcp-snooping	1	FastEthernet0/4

Total number of bindings: 1

把 R1 的 G0/0 端口关闭，采用相同方式测试 R2 和 R4 能否从 R3 获得 IP 地址。正确结果应该是，R2 和 R4 无法从 R3 获得 IP 地址，即用户私自搭建的 DHCP 服务器无法提供 IP 地址。

提示

DHCP 欺骗防范通常在接入层交换机上采用（例如以上的交换机 S2）。不过遗憾的是，作为接入层交换机使用的早期 Catalyst 2950 交换机并不支持 DHCP Snooping 功能。

（5）option 82 问题

这个选项被称为 DHCP Relay Agent Information Option（中继代理信息选项），选项号为 82，故又称为 option 82。当 DHCP 服务器和客户端不在同一个子网内时，客户端要想从 DHCP 服务器上分配到 IP 地址，就必须由 DHCP 中继代理（DHCP Relay Agent）来转发 DHCP 请求包。DHCP 中继代理在将客户端的 DHCP 报文转发到 DHCP 服务器之前，可以插入一些选项信息，以便 DHCP 服务器能更精确地得知客户端的信息。

开启了 DHCP 监听功能后，默认情况下，交换机将对从非信任端口接收到的 DHCP 请求报文插入选项 82 信息。在本实验中，由于 DHCP Server 和 DHCP Client 在同一 VLAN 中，交换机把 option 82 的值填为 0，而这将导致基于 IOS 的路由器 R1 无法正常分配 IP 地址。因此在上面的步骤中，交换机 S1 和 S2 上都配置了"**no ip dhcp snooping information option**"。但是如果把 R1 换成 Windows Server 2008、2012 服务器，则无须此命令，因为 Windows Server 2008、2012 的 DHCP 服务器不认为 option 82 的值为 0 的 DHCP 报文是错误的。

有另一方法也可以解决 option 82 问题，如下所示：

```
S2(config)#ip dhcp snooping information option
S1(config)#ip dhcp snooping information option
//保留交换机 S1 和 S2 添加 Option 82 选项的功能

S1(config)#interface FastEthernet0/23
S1(config-if)#ip dhcp snooping trust
S1(config-if)#interface FastEthernet0/24
S1(config-if)#ip dhcp snooping trust
R1(config)#ip dhcp relay information trust-all
//配置基于 IOS 的 DHCP Server，使其能够接收 Option 82 为 0 的 DHCP 报文。
```

由于交换机 S2 会在 DHCP 报文中添加 option 82 选项，当 S2 把 DHCP 发往 S1 时，如果 S1 的 Fa0/23 和 Fa0/24 端口不是信任端口，交换机 S1 将拒绝接收，因此需要把这两个端口配置为信任端口。要注意的是，两个端口都需要配置，因为 DHCP 报文都有可能从这两条链路进行发送。还可以采用以下方式来替代：

```
S1(config)#ip dhcp snooping information option allow-untrusted
//配置交换机 S1，如果从非信任端口接收到的 DHCP 报文中带有 option 82 选项，也接收该 DHCP 报文
```

采用步骤（4）的测试方式进行测试，R2 和 R4 应该能够正常获得 IP 地址。

说明

限于篇幅，无法进一步介绍 option 82 选项插入导致的其他问题，请参阅思科网站的资料。

（6）显示 DHCP 监听信息

```
S1#show ip dhcp snooping
Switch DHCP snooping is enabled    //启用 DHCP 监听
DHCP snooping is configured on following VLANs:   //配置监听的 VLAN
1
DHCP snooping is operational on following VLANs:   //实际监听的 VLAN
1
DHCP snooping is configured on the following L3 Interfaces:
//配置监听的三层端口，本实验没有配置

Insertion of option 82 is enabled    //启用 option 82 插入功能
    circuit-id format: vlan-mod-port
    remote-id: d0c7.89ab.1180 (MAC)
Option 82 on untrusted port is allowed    //允许从非信任端口接收带 option 82 的 DHCP 报文
Verification of hwaddr field is enabled
//检查 DHCP 报文中的 MAC 地址和以太网帧的 MAC 地址是否相同
Verification of giaddr field is enabled
DHCP snooping trust/rate is configured on the following Interfaces:
Interface                Trusted      Rate limit (pps)
-----------------------  -------      ----------------
FastEthernet0/1          yes          unlimited
//以上显示了信任端口，以及端口的 DHCP 报文数量限制
```

Interface	Trusted	Allow option	Rate limit (pps)
FastEthernet0/23	yes	yes	unlimited
Custom circuit-ids:			
FastEthernet0/24	yes	yes	unlimited
Custom circuit-ids:			

S1#**show ip dhcp snooping binding**

MacAddress	IpAddress	Lease(sec)	Type	VLAN	Interface
F8:72:EA:69:1C:78	172.16.1.9	86256	dhcp-snooping	1	FastEthernet0/2
F8:72:EA:C8:4F:99	172.16.1.7	85982	dhcp-snooping	1	FastEthernet0/23

Total number of bindings: 2

各列的含义如下所述。

① "MacAddress"：DHCP Client 的 MAC 地址。
② "IpAddress"：DHCP Client 的 IP 地址。
③ "Lease(sec)"：IP 地址的租约时间（秒）。
④ "Type"：记录的类型，dhcp-snooping 表明是动态生成的记录。
⑤ "VLAN"：VLAN 编号。
⑥ "Interface"：端口。

S1#**show ip dhcp snooping database**
Agent URL : flash:dhcp_snooping.db   //数据库的 URL
Write delay Timer : 15 seconds   //写入数据库的延迟时间
Abort Timer : 15 seconds
Agent Running : No
Delay Timer Expiry : Not Running
Abort Timer Expiry : Not Running
Last Succeded Time : 01:41:09 UTC Mon Mar 1 1993   //上次成功写入的时间
Last Failed Time : None   //上次写入失败的时间
Last Failed Reason : No failure recorded.   //写入失败的原因

Total Attempts	:	2	Startup Failures :	0
Successful Transfers	:	2	Failed Transfers :	0
Successful Reads	:	0	Failed Reads :	0
Successful Writes	:	2	Failed Writes :	0
Media Failures	:	0		

//各动作的统计数

6.5 实验 4：DAI 与 IPSG

1. 实验目的

通过本实验可以掌握：

① DAI（Dynamic ARP Inspection）的原理。
② DAI 的配置。
③ IP Source Guard 的原理。
④ IP Source Guard 的配置。

2. 实验拓扑

实验拓扑如图 6-9 所示。由于 DAI 和 IPSG 与 IP DHCP Snooping 的紧密关系，这里将在 6.4 节的基础上配置 DAI 和 IPSG。

3. 实验步骤

在 6.4 节中，交换机 S1 和 S2 将通过 DHCP 监听学习到各端口上的绑定信息，如下所示：

```
S1#show ip dhcp snooping binding
MacAddress          IpAddress       Lease(sec)    Type           VLAN    Interface
------------------  --------------  -----------   -------------  ----
F8:72:EA:69:1C:78   172.16.1.12     86326         dhcp-snooping  1       FastEthernet0/2
F8:72:EA:C8:4F:99   172.16.1.13     86391         dhcp-snooping  1       FastEthernet0/23
Total number of bindings: 2
```
//从交换机 S1 上看到了 DHCP 监听的绑定信息。接在 Fa0/23 上的计算机（实际上就是 R4）的 MAC 地址为 F8:72:EA:C8:4F:99，IP 地址为 172.16.1.13 ；接在 Fa0/2 上的计算机（实际上就是 R2）的 MAC 地址为 F8:72:EA:69:1C:78，IP 地址为 172.16.1.12

```
S2#show ip dhcp snooping binding
MacAddress          IpAddress       Lease(sec)    Type           VLAN    Interface
------------------  --------------  -----------   -------------  ----
F8:72:EA:C8:4F:99   172.16.1.13     86343         dhcp-snooping  1       FastEthernet0/4
Total number of bindings: 1
```

（1）在 S1 上配置 DAI

```
S1(config)#ip arp inspection vlan 1
```
//在 VLAN1 中启用 DAI
```
S1(config)#ip arp inspection validate src-mac dst-mac ip
```
配置 DAI 要检查 ARP 报文（包括请求和响应）中的源 MAC 地址、目的 MAC 地址、源 IP 地址和 DHCP Snooping 绑定中的信息是否一致。

S1 的 Fa0/1 端口接的是可信任的路由器 R1，因此该端口可以配置为 DAI 信任端口，其他端口则默认为 DAI 非信任端口。DAI 将不对 Fa0/1 的 ARP 包进行检测。配置如下：

```
S1(config)#interface FastEthernet0/1
S1(config-if)#ip arp inspection trust
```

S1 的 Fa0/23 和 Fa0/24 是下连到 S2 的端口，这两个端口可能会收到大量的 ARP 包。因此取消 ARP 包的限制，默认值是 15 包/秒。

```
S1(config)#interface FastEthernet0/23
S1(config-if)#ip arp inspection limit none
```

测试从 R2 上能否 ping 通 R1,操作如下:

```
R2#ping 172.16.1.1    //172.16.1.1 是 R1 的 IP 地址
Type escape sequence to abort.
Sending 5, 100-byte ICMP Echos to 172.16.1.1, timeout is 2 seconds:
!!!!!
Success rate is 100 percent (5/5), round-trip min/avg/max = 1/2/4 ms
```

(2)在 S2 上配置 DAI

```
S2(config)#ip arp inspection vlan 1
S2(config)#ip arp inspection validate src-mac dst-mac ip
```

这样配置后,R4 能否与 R1 和 R2 通信?答案是不行的。当 R4 发送 ARP 请求查询 R1 和 R2 的 MAC 地址时,S1 和 S2 的 DHCP 监听绑定表中有 R4 的信息,允许 R4 的 ARP 请求包通过。然而当 R1 响应 MAC 地址查询请求时,交换机 S2 的 DHCP 监听绑定表中并没有 R1 和 R2 的信息,因此将拒绝 ARP 响应包。因此需要配置如下:

```
S2(config)#interface FastEthernet0/23
S2(config-if)#ip arp inspection trust
S2(config)#interface FastEthernet0/24
S2(config-if)#ip arp inspection trust
//配置 Fa0/23 和 Fa0/24 为信任端口,将不对这些端口收到的 ARP 包进行检查

R4#ping 172.16.1.1    //172.16.1.1 是 R1 的 IP 地址
Type escape sequence to abort.
Sending 5, 100-byte ICMP Echos to 172.16.1.1, timeout is 2 seconds:
!!!!!
Success rate is 100 percent (5/5), round-trip min/avg/max = 1/2/4 ms

R4#ping 172.16.1.12    //172.16.1.12 是 R2 的 IP 地址
Type escape sequence to abort.
Sending 5, 100-byte ICMP Echos to 172.16.1.11, timeout is 2 seconds:
!!!!!
Success rate is 100 percent (5/5), round-trip min/avg/max = 1/1/4 ms
```

提示

可以把黑客不可能接触到的端口配置为信任端口,例如,本实验中的 S1 和 S2 的 Fa0/23 和 Fa0/24 端口为级联端口,并不对用户开放。这样做的好处是减少复杂性,否则得详细考虑哪些端口需要放行 ARP 包。

(3)计算机使用静态 IP 地址时配置 DAI

假设本实验拓扑中 R3 因某种原因需要使用固定 IP 地址 172.16.1.3 。有两个方案,方案一:在 DHCP 服务器上为 R3 分配固定的 IP 地址,如下所示。

```
R1(config)#ip dhcp pool HOST-R3
R1(dhcp-config)#host 172.16.1.3 255.255.255.0
R1(dhcp-config)#client-identifier 0100.f872.ea69.18b9
```

//创建一个地址池，实际上就一个 IP 地址。该地址固定分配给 ID 为 0100．f872.ea69.18b9 的 DHCP 客户。ID 的组成为 0100+DHCP 客户以太网端口的 MAC 地址

R3(config)#**interface GigabitEthernet0/1**
R3(config-if)#**ip dhcp client client-id GigabitEthernet0/1**
//配置 R3 把 G0/1 端口的 MAC 地址作为 Client-id 。
R3(config-if)#**ip address dhcp**
R3(config-if)#**no shutdown**
*Feb 5 01:57:54.938: %DHCP-6-ADDRESS_ASSIGN: Interface GigabitEthernet0/1 assigned DHCP address 172.16.1.14, mask 255.255.255.0, hostname R3
//R3 的 G0/1 端口还是配置为动态获得 IP 地址

提示

在 R1 的 DHCP 服务器上配置固定 IP 时，需要 DHCP 客户的 ID。如果难于确定 DHCP 客户的 ID，可以先让 DHCP 客户端通过 DHCP 获得一个 IP 地址，然后在 R1 上使用 "**show ip dhcp binding**" 命令查看客户端的 ID，再使用该 ID 配置分配固定的 IP 地址。如果 DHCP Server 是 Windows Server 或者 Linux 操作系统，请参阅相关资料为 DHCP Client 分配固定 IP 地址。

当采用方案一时，由于 R3 的 IP 地址还是通过 DHCP 获得的，因此 DHCP Snooping 还是能够检测到的，在 DHCP Snooping binding 表中有该记录。因此基于 DHCP Snooping binding 记录工作的 DAI 是能正常工作的。

```
S1#show ip dhcp snooping binding
MacAddress          IpAddress        Lease(sec)    Type            VLAN     Interface
------------------  ---------------  ------------  --------------  ----     --------------
F8:72:EA:C8:4F:99   172.16.1.13      85822         dhcp-snooping   1        FastEthernet0/23
(略去了不相关的信息)

S2#show ip dhcp snooping binding
MacAddress          IpAddress        Lease(sec)    Type            VLAN     Interface
------------------  ---------------  ------------  --------------  ----     --------------
F8:72:EA:C8:4F:99   172.16.1.13      85802         dhcp-snooping   1        FastEthernet0/4
(略去了不相关的信息)
```

方案二：直接在 R3 上使用静态 IP 地址。

R3(config)#**interface GigabitEthernet0/1**
R3(config-if)#**no ip address**   //先释放动态获得的 IP 地址
R3(config-if)#**ip address 172.16.1.3 255.255.255.0**

这时 S1 和 S2 的 DHCP Snooping binding 表中将没有该记录，R3 发送的 ARP 包将不被交换机接收。我们可以手工添加，如下所示：

S2#**ip dhcp snooping binding f872.ea69.18b9 vlan 1 172.16.1.3 interface fastethernet0/3 expiry 4294967295**
S1#**ip dhcp snooping binding f872.ea69.18b9 vlan 1 172.16.1.3 interface fastethernet0/23 expiry 4294967295**

f872.ea69.18b9 是 R3 的 MAC 地址，172.16.1.3 是 R3 的 IP 地址，4294967295 表示绑定永不过期。需要注意的是：以上命令是直接在特权模式下执行的，如果 ARP 通过 Fa0/24 端

口接收，就得添加到 Fa0/24 端口，因此可能会给管理员带来困难。

```
S1#show ip dhcp snooping binding
MacAddress          IpAddress       Lease(sec)   Type           VLAN   Interface
------------------  --------------  ------------ -------------- ----   --------------
F8:72:EA:69:18:B9   172.16.1.3      infinite     dhcp-snooping  1      FastEthernet0/23
(略去了不相关的信息)

S2#show ip dhcp snooping binding
MacAddress          IpAddress       Lease(sec)   Type           VLAN   Interface
------------------  --------------  ------------ -------------- ----   --------------
F8:72:EA:69:18:B9   172.16.1.3      infinite     dhcp-snooping  1      FastEthernet0/3
(略去了不相关的信息)
```

手工添加 DHCP Snooping 绑定后，基于 DHCP Snooping binding 记录工作的 DAI 就能正常工作了。R1、R2、R3 和 R4 应该能互相通信。

（4）查看 DAI 的信息

```
S1#show ip arp inspection
Source Mac Validation      : Enabled     //启用了源 MAC 地址检查
Destination Mac Validation : Enabled     //启用了目的 MAC 地址检查
IP Address Validation      : Enabled     //启用了源 IP 地址检查

  Vlan     Configuration    Operation    ACL Match          Static ACL
  ----     -------------    ---------    ---------          ----------
   1       Enabled          Active

  Vlan     ACL Logging      DHCP Logging     Probe Logging
  ----     -----------      ------------     -------------
   1       Deny             Deny             Off

  Vlan     Forwarded        Dropped      DHCP Drops     ACL Drops
  ----     ---------        -------      ----------     ---------
   1       166              5            5              0

  Vlan     DHCP Permits     ACL Permits  Probe Permits  Source MAC Failures
  ----     ------------     -----------  -------------  -------------------
   1       104              0            0              0

  Vlan     Dest MAC Failures   IP Validation Failures   Invalid Protocol Data
  ----     -----------------   ----------------------   ---------------------

  Vlan     Dest MAC Failures   IP Validation Failures   Invalid Protocol Data
  ----     -----------------   ----------------------   ---------------------
   1       0                   0                        0

S1#show ip arp inspection interfaces
Interface       Trust State    Rate (pps)    Burst Interval
---------       -----------    ----------    --------------
Fa0/1           Trusted        None          N/A
Fa0/2           Untrusted      15            1
Fa0/3           Untrusted      15            1
```

(略去了部分信息)
//以上显示了端口是否可信任及 ARP 包数量的限制

(5) 在 S1 上配置 IPSG

默认时,IPSG 不检查所有端口的数据包,因此 IPSG 只需要在不信任端口上进行配置。如下所示:

S1(config)# **interface fastethernet0/2**
S1(config-if)#**ip verify source port-security**
//在 Fa0/2 端口启用 IPSG 功能。不加 port-security 参数,表示 IP 源防护功能只执行"源 IP 地址过滤"模式;加 port-security 参数,表示 IP 源防护功能执行"源 IP 地址和源 MAC 地址过滤"模式
S1(config-if)#**switchport mode access**
S1(config-if)#**switchport port-security**
S1(config-if)#**switchport port-security mac-address sticky**
S1(config-if)#**switchport port-security violation shutdown**
//由于开启了"源 IP 地址和源 MAC 地址过滤",所以以上同时配置端口安全功能

技术要点

IPSG 有两种过滤类型:源 IP 地址过滤、源 IP 地址和源 MAC 地址过滤。源 IP 地址过滤——根据源 IP 地址对 IP 流量进行过滤,只有当源 IP 地址与 IP 源绑定条目匹配时 IP 流量才允许通过。源 IP 地址和源 MAC 地址过滤——根据源 IP 地址和源 MAC 地址对 IP 流量进行过滤,只有当源 IP 地址和源 MAC 地址都与 IP 源绑定条目匹配时 IP 流量才允许通过。

当交换机只使用"IP 源地址过滤"时,IP 源防护功能与端口安全功能是相互独立的关系。而当交换机使用"源 IP 功能和源 MAC 地址过滤"功能时,IP 源防护功能与端口安全功能是"集成"关系,更确切地说需要同时开启端口安全功能才能完成源 IP 地址和源 MAC 地址过滤。

当以源 IP 地址和源 MAC 地址作为过滤条件时,为了确保 DHCP 协议能够正常地工作,还必须启用 DHCP 监听选项 82。对于没有选项 82 的数据,交换机不能确定用于转发 DHCP 服务器响应的客户端主机端口,DHCP 服务器响应将被丢弃,客户机也不能获得 IP 地址。

```
S1#show ip verify source
Interface  Filter-type  Filter-mode  IP-address    Mac-address        Vlan  Log
---------  -----------  -----------  -----------   -----------        ----  ---
Fa0/2      ip-mac       active       172.16.1.12   F8:72:EA:69:1C:78  1     disabled
```
//以上显示 IPSG 在 Fa0/2 端口上检查数据包的源 IP 地址和目的 IP 地址,只有符合条件的数据包才被放行

```
S1#show ip source binding
MacAddress         IpAddress     Lease(sec)  Type            VLAN  Interface
-----------------  ------------  ----------  --------------  ----  ----------------
F8:72:EA:69:18:B9  172.16.1.3    infinite    dhcp-snooping   1     FastEthernet0/23
F8:72:EA:69:1C:78  172.16.1.12   85196       dhcp-snooping   1     FastEthernet0/2
F8:72:EA:C8:4F:99  172.16.1.13   85261       dhcp-snooping   1     FastEthernet0/23
Total number of bindings: 3
```

(6) 在 S2 上配置 IPSG

S2(config)#**interface range FastEthernet0/3 -4**

S2(config-if)# **switchport mode access**
S2(config-if)# **switchport port-security**

S1(config)#**interface FastEthernet0/2**
S1(config-if)#**switchport port-security max 1**
S1(config-if)#**switchport port-security violation shutdown**
S2(config-if)# **switchport port-security mac-address sticky**
S2(config-if)# **ip verify source port-security**

（7）用户计算机使用静态 IP 地址时 IPSG 的配置

还是以 R3 需要使用固定 IP 地址为例。和步骤（3）中的两个方案类似，如果采用方案一，则 IPSG 无须特殊配置；如果采用方案二，即 R3 使用静态 IP 地址，则配置如下：

R3(config)#**interface GigabitEthernet0/1**
R3(config-if)#**no ip address**
R3(config-if)#**ip address 172.16.1.3 255.255.255.0**
S2 (config)#**ip source binding f872.ea69.18b9 vlan 1 172.16.1.3 interface Fa0/3**
// f872.ea69.18b9 是 R3 的 MAC 地址，172.16.1.3 是 R3 的 IP 地址，永不过期。
S2#**show ip source binding**

MacAddress	IpAddress	Lease(sec)	Type	VLAN	Interface
F8:72:EA:69:18:B9	172.16.1.3	infinite	static	1	**FastEthernet0/3**
F8:72:EA:C8:4F:99	172.16.1.13	84929	dhcp-snooping	1	FastEthernet0/4

Total number of bindings: 2
//手工添加了一条记录

提示

当使用 "S2(config)#**ip source binding f872.ea69.18b9 vlan 1 172.16.1.3 interface Fa0/3**" 命令时，实际上也会自动在 "ip source binding" 表中创建记录。然而使用 "ip source binding" 命令创建记录，在 "ip dhcp shooping binding" 表中并不会自动创建记录。

技术要点

如果有客户端使用固定 IP 地址，建议还是通过配置 DHCP 服务器为客户分配固定 IP 地址，而不是在客户端上直接配置静态 IP 地址。因为前者在 DHCP 监听表中有记录，基于该表工作的 DAI 和 IPSG 能正常运行，不需要手工添加相关静态记录。

6.6　实验 5：AAA

1. 实验目的

通过本实验可以掌握：
① Cisco ACS 的安装和配置。
② 如何在交换机上配置 AAA。

2. 实验拓扑

AAA 实验拓扑如图 6-10 所示。下一实验也将使用该拓扑，本实验不使用路由器 R1。本实验将在 ACS Server 上安装 Cisco 的 ACS 软件，在交换机 S1 上配置使用 AAA，在 Test PC 上进行登录测试。

  提示

> 思科已经在 2017 年年底停止销售 ACS 软件，最新的版本是 5.8 版，5.8 版在 Linux 系统上安装。本书重点是理解 AAA，而不是 ACS 服务器，因此为简单起见还使用 Windows Server 2003 上安装的 ACS v4.0，虽然有点老，但并不妨碍对 AAA 的讲解。作者已经把装好的 VMware 虚拟机上传到网盘以减少读者的工作量，链接为 https://pan.baidu.com/s/1kWoIZxt 密码：s22n。

3. 实验步骤

（1）ACS 软件的安装

Cisco Secure ACS（Access Control Server）是思科公司推出的 AAA 服务器软件，可以安装在 Windows Server 上。本实验使用 CiscoSecure ACS v4.0，安装该软件要求先安装 Java 运行环境：Sun JRE 1.4.2_04 或更高版。ACS 软件的安装步骤如下所述。

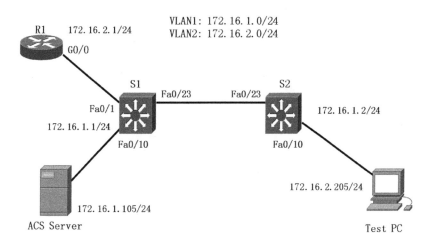

图 6-10　AAA 实验拓扑

运行安装程序 setup.exe 后，直接单击"Accept"按钮，再单击"Next"按钮。如图 6-11 所示，选中全部选项表示系统已经满足了全部的安装要求，单击"Next"按钮两次。

如图 6-12 所示，保持默认选项，AAA 将使用 ACS 内部的数据库作为用户数据库，而不是使用 Windows Server 的用户数据库，单击"Next"按钮开始安装。

如图 6-13 所示，可以选择 ACS 管理界面中要显示的一些选项，这里先不选，稍后再配置。单击"Next"按钮。

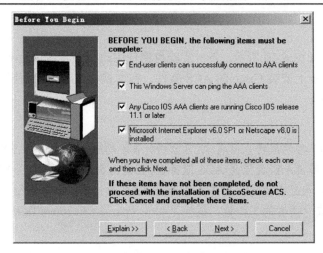

图 6-11　安装 ACS 需要满足的条件

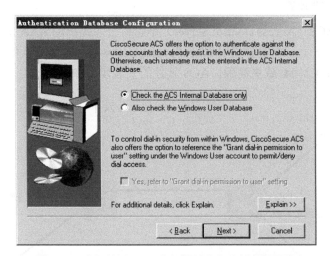

图 6-12　选择使用 ACS 内部数据库作为用户数据库

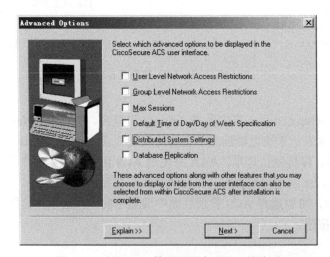

图 6-13　选择 ACS 管理界面中要显示的选项

第 6 章 交换机的安全

如图 6-14 所示，可以选择 ACS 故障后是否重启软件以及发送通知邮件，这里保持默认值。单击"Next"按钮。

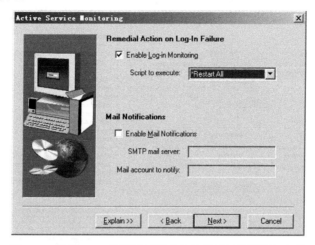

图 6-14 选择 ACS 故障后是否重启软件以及发送通知邮件

如图 6-15 所示，输入加密数据库的密码，请勿丢失该密码。单击"Next"按钮两次，完成安装。桌面上会有一个"ACS Admin"的快捷方式，双击打开浏览器，如图 6-15 所示是 ACS 的管理界面。

图 6-15 ACS 管理的界面

（2）配置认证

通常管理员是 Telnet 到交换机来配置交换机的，这里假设需要为不同管理员创建不同的用户。使用 AAA 认证后，管理员 Telnet 到交换机后输入用户名和密码，交换机会作为 AAA 的客户端在 AAA 服务器上进行认证并允许或者不允许用户 Telnet。

要在 ACS 上创建 AAA Client，先打开 ACS 的管理界面，单击"Network Configuration"，

如图 6-16 所示，单击"AAA Clients 区"的"Add Entry"，添加 AAA Client。

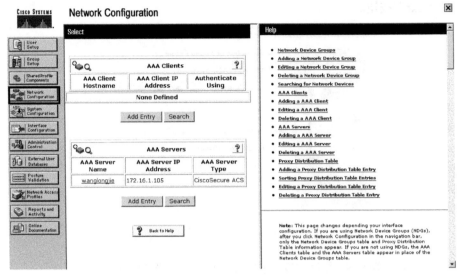

图 6-16　添加 AAA Client

如图 6-17 所示，填写交换机 S1 的 AAA Client 信息，其中，S1 是交换机上的 hostname；172.16.1.1 是交换机的 IP 地址；Key 是 S1 和 ACS 之间相互验证的密码。此外，这里选择 TACACS 服务器来进行认证。单击"Submit+Apply"提交并生效。

图 6-17　AAA Client 信息

下面要创建用户，如图 6-18 所示，单击"User Setup"，输入要创建的用户名，单击"Add/Edit"按钮。

第 6 章 交换机的安全

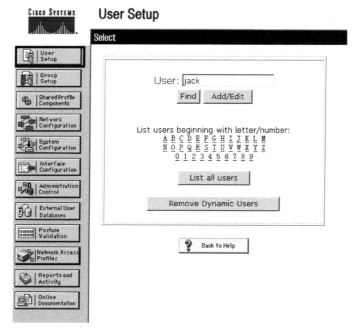

图 6-18 创建用户

如图 6-19 所示，输入用户的信息，单击"Submit"提交。

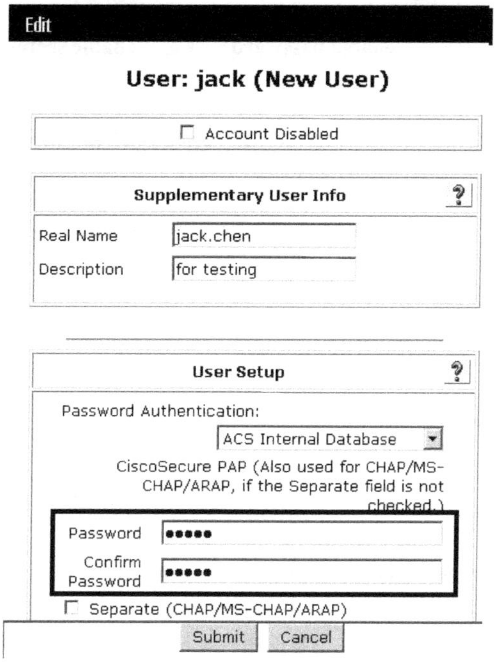

图 6-19 输入用户信息

交换机 S1 配置如下：
Switch(config)#**hostname S1**
S1(config)#**interface vlan 1**
S1(config-if)#**no shutdown**

S1(config-if)#**ip address 172.16.1.1 255.255.255.0**
S1(config)#**aaa new-model**   //启用 AAA 认证功能
S1(config)#**tacacs-server host 172.16.1.105**
//指明 AAA 服务器的 IP 地址
S1(config)# **tacacs-servertac   key cisco**
//指明在和 AAA 服务器相互验证时的密码

S1#**test aaa group tacacs jack cisco legacy**
Attempting authentication test to server-group tacacs+ using tacacs+
User was successfully authenticated.
//测试刚创建的用户是否正常。jack 是用户名，cisco 是该用户的密码
S1(config)#**aaa authentication login TEST_LOGIN group tacacs**
//创建一个认证列表，该列表使用 tacacs 进行认证

S1(config)#**line vty 0 5**
S1(config-line)#**login authentication TEST_LOGIN**
//配置使用之前创建的认证列表对 Telnet 或者 SSH 用户进行认证

S1(config)#**aaa authentication enable default enable**
S1(config)#**enable secret cisco**
//用户 Telnet 到交换机后，还需要执行"enable"命令才能配置交换机，需要 enable 密码。这里配置使用交换机本地的 enable 密码，即使用"**enable password**"或者"**enable secret**"命令所设置的密码

 提示

从管理上来讲，应该为不同用户设置不用的 enable 密码。然而 ACS 并不支持，这里使用了交换机本地的 enable 密码。

在交换机 S1 上配置 VLAN 间路由，保证 Test PC 能 ping 通 S1。从 Test PC 上 Telnet 交换机 S1：

C:\>**telnet 172.16.1.1**
User Access Verification
Username: jack
Password: //用户的密码是 cisco
S1>**enable**
Password: //密码是 cisco
S1#**conf terminal**
Enter configuration commands, one per line.   End with CNTL/Z.
S1(config)#
//以上说明用户能够使用 AAA 服务器上创建的用户正常地 Telnet 到交换机上

（3）配置授权

在步骤（2）中，用户成功 Telnet 到交换机后，就可以为所欲为了。可以配置授权，明确指明用户可以执行什么命令。这里假设要授权 jack 用户可以进行以下操作：

S1#**configure termina**
S1(config)#**ip routing l**

S1(config)#**router rip**
S1(config-router)#**version 2**
S1(config-router)#**network 172.16.0.0**

可以为每个用户单独进行授权，也可以为组进行授权，这里选择后者。jack 用户默认属于 Default Group 组。如图 6-20 所示，当选择组后，单击"Edit Settings"按钮。

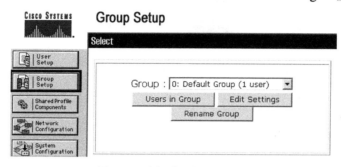

图 6-20　选择默认组进行编辑

如图 6-21 所示，选中"Shell(exec)"，授权组能够进入 exec 模式（即特权模式#）。

如图 6-22 所示进行配置，"Unmatched Cisco IOS commands"为 deny，意思是凡是没有明确授权的命令不能执行。在图 6-22 中配置允许执行 configure 命令，可以带参数 terminal，不能带其他参数。单击"Submit + Restart"按钮提交。

图 6-21　授权组有 exec 权限

如图 6-23 所示配置其他命令的授权，注意：network 命令和 version 命令应该可以带任何参数，所以是 permit 未指明的参数。

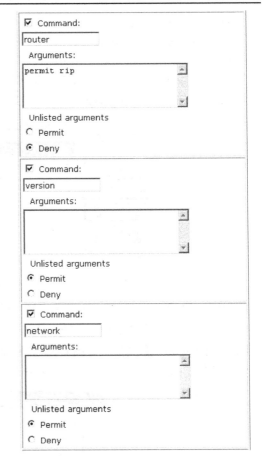

图 6-22 授权组可以执行命令 configure　　图 6-23 授权组可以执行的其他命令

交换机配置如下：

S1(config)#**aaa authorization exec TEST group tacacs+**
//配置 exec 模式的授权列表，exec 模式就是特权模式"#"

S1(config)#**aaa authorization config-commands**
//要求配置模式下的命令也要授权，配置模式就是"S1(config)#"模式

S1(config)#**aaa authorization configuration TEST group tacacs+**
//配置 config 模式的授权列表

S1(config)#**aaa authorization commands 15 TEST group tacacs+**
//配置命令等级为 15 的命令授权列表。

提示

在 IOS 中，不同命令有不同的默认等级，例如，**disable** 的等级是 1，而 **show running**、**configure terminal** 和 **router rip** 等命令的等级是 15。用户登录后也有不同的默认等级，用户模式下是 1，特权模式下是 15，用户可以执行比自己等级低的命令。使用 **show privi** 命令可以显示自己的等级。

S1(config)#**line vty 0 5**
S1(config-line)#**authorization exec TEST**
S1(config-line)#**authorization commands 15 TEST**

在配置用户 Telnet 后，在 exec 模式下，执行等级为 15 的命令使用列表 Test 进行授权。以上只要求等级为 15 的命令要授权，这意味着等级为其他的命令就不需要授权了，因此用户可以不经授权执行"show version"命令。

从 Test-PC 上 Telnet 交换机进行测试：

```
S1#show running-config
Command authorization failed.
// show running-config 命令并没有被授权
S1#configure terminal
Enter configuration commands, one per line.    End with CNTL/Z.
S1(config)#router ospf 1
Command authorization failed.
//router 命名不允许带 ospf 参数
S1(config)#router rip
S1(config-router)#version 2
S1(config-router)#network 172.16.0.0
S1(config-router)#passive-interface vlan 1
Command authorization failed.
// passive-interface vlan 1 命令并没有被授权
```

（4）配置审计

审计就是记录用户所做的事情。

```
S1(config)#aaa accounting exec ACC start-stop group tacacs+
S1(config)#aaa accounting commands 0 ACC start-stop group tacacs+
S1(config)#aaa accounting commands 1 ACC start-stop group tacacs+
S1(config)#aaa accounting commands 15 ACC start-stop group tacacs+
//配置用户进入 exec 模式，命令等级为 0、1、15 的审计列表
S1(config)#line vty 0 5
S1(config-line)#accounting exec ACC
S1(config-line)#accounting commands 0 ACC
S1(config-line)#accounting commands 1 ACC
S1(config-line)#accounting commands 15 ACC
//配置 Telnet 用户进入 exec 模式，执行等级为 0、1、15 的命令都要审计
```

从 Test-PC 上 Telnet 交换机进行测试：

```
S1>enable
S1#configure terminal
S1(config)#router rip
S1(config-router)#version 2
S1(config-router)#network 172.16.0.0
```

用户登录报告如图 6-24 所示，单击"Reports and Activity"→"TACACS+ Accounting"，可以看到用户登录和注销的时间。

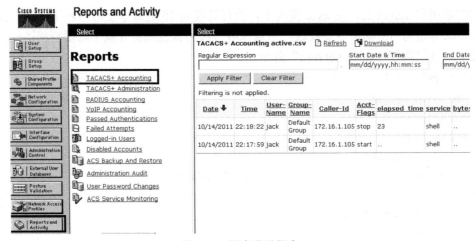

图 6-24 用户登录报告

如图 6-25 所示，单击 "Reports and Activity" → "TACACS+ Administration"，可以看到用户执行的命令。

图 6-25 用户执行过的命令

6.7 实验 6：dot1x

1. 实验目的

通过本实验可以掌握：
① 在交换机上配置 dot1x 的方法。
② 使用 dot1x 将计算机连接到网络的方法。

2. 实验拓扑

实验拓扑如图 6-10 所示。本实验将在交换机 S2 上配置 dot.1x，在 Test PC 上进行测试。当用户输入正确的用户名和密码后，交换机 S2 把 Test PC 划分到 VLAN2 中。

3. 实验步骤

（1）在 ACS 上创建 AAA 客户

只有 RADIUS 才支持 dot1x，因此在 ACS 上添加 S2 作为 RADIUS 客户。在 ACS 的管理界面单击"Network Configuration"→"Add Entry"，AAA Client 的信息如图 6-26 所示，输入 S2 的相关信息后，单击"Submit＋Apply"提交。

图 6-26　AAA Client 的信息

单击"Group Setup"，选择 jack 用户所在的组"Default Group"，单击"Edit Setting"按钮，如图 6-27 所示设置[064]、[065]和[081]。[081]选项是 jack 用户登录后所在的 VLAN。

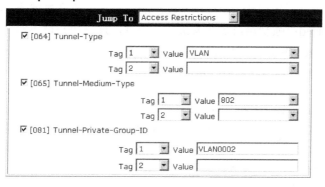

图 6-27　配置用户登录后的 VLAN 信息

提示

应该事先在交换机 S1 和 S2 上创建 VLAN2，并在交换机 S1 上配置 VLAN 间的路由功能。

(2) 在交换机上配置 dot1x

交换机 S2 配置如下:

```
Switch(config)#hostname S2
S2(config)#interface vlan 1
S2(config-if)#ip address 172.16.1.2 255.255.255.0
S2(config-if)#exit
S2(config)#aaa new-model
S2(config)#radius-server host 172.16.1.105
//配置 AAA 服务器的地址
S2(config)#radius-server key cisco
//配置 AAA 服务器的密码
S2(config)#aaa authentication dot1x default group radius
//配置 dot1x 使用 RADIUS 服务器
S2(config)#aaa authorization network default group radius
//定义 dot1x 授权策略
S2(config)#dot1x system-auth-control
//全局开启 dot1x 控制
S2(config)#radius-server vsa send authentication
//配置交换机发送厂商特别属性到 AAA 服务器,目的是要获取用户的 VLAN 信息
S2#test aaa group radius jack cisco legacy
Attempting authentication test to server-group radius using radius
User was successfully authenticated.
//测试 AAA 服务是否正常


S2(config)#interface FastEthernet0/10
S2(config-if)#switchport mode access
S2(config-if)#authentication port-control auto
```

authentication port-control 可以带以下 3 个参数中的之一。

① auto:认证通过后端口状态就变为 force-authorized,不通过就为 force-unauthorized。

② force-authorized:强制端口状态为认证已通过,这样用户就不需要认证了。

③ force-unauthorized:强制端口状态为不认证通过,这样用户实际上就不能使用端口了。

```
S2(config-if)#authentication host-mode single-host
//该端口只能是一台计算机接入,如果端口接的是一个 Hub 并接有多台计算机,可以使用 multi-host
S2(config-if)#authentication violation protect
//认证不通过(违规)时端口的动作,有 shutdown、restrict、protect;shutdown 是关闭端口,protect 会在日志中记录,restrict 不在日志中记录
S2(config-if)#dot1x pae authenticator
//设置该端口为 IEEE 802.1x 认证
S2(config-if)#spanning-tree portfast
S2(config-if)#exit
S2(config)#dot1x reauthentication
//认证失败,重新认证
```

（3）测试

在 Test PC（Windows 7）上进行测试。首先在 Test PC 上把以下文本保存到 1.reg 文件中，双击 1.reg 文件，把注册表项导入到注册表中。

```
Windows Registry Editor Version 5.00
[HKEY_LOCAL_MACHINE\SYSTEM\CurrentControlSet\services\RasMan\PPP\EAP\4]
"FriendlyName"="MD5-Challenge"
"RolesSupported"=dword:0000000a
"Path"=hex(2):25,00,53,00,79,00,73,00,74,00,65,00,6d,00,52,00,6f,00,6f,00,74,\
  00,25,00,5c,00,53,00,79,00,73,00,74,00,65,00,6d,00,33,00,32,00,5c,00,52,00,\
  61,00,73,00,63,00,68,00,61,00,70,00,2e,00,64,00,6c,00,6c,00,00,00
"InvokeUsernameDialog"=dword:00000001
"InvokenPasswordDialog"=dword:00000001
```

如图 6-28 所示，在 Test PC 上首先从 "计算机管理"→"服务和应用程序"→ "服务" 启动 Wired AutoConfig 服务。

图 6-28　启用 Wired AutoConfig 服务

如图 6-29 所示，在网络本地连接属性中，启用 "IEEE 802.1X 身份验证"，网络身份验证方法为 "MD5-Challenge"。

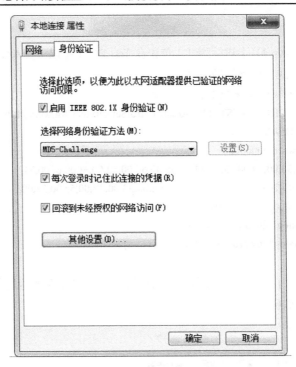

图 6-29　启用"IEEE 802.1x 身份验证"

如图 6-30，设置"IEEE 802.1X 身份验证模式"，使用"用户身份验证"。

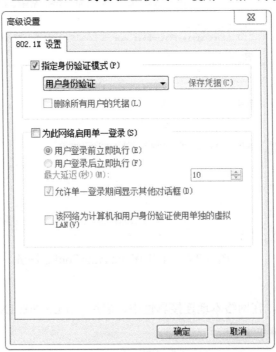

图 6-30　设置"IEEE 802.1X 身份验证模式"

把网卡禁用并重启，如图 6-31 所示，在 Windows 窗口右下角出现用户名和密码提示。双击图标，如图 6-31 所示输入用户名和密码。

图 6-31 输入用户名和密码

在 Test PC 上测试，因为这里把 Test-PC 划分到 VLAN2 中，所以 IP 地址改为 172.16.2.205。

C:\>**ipconfig**
Ethernet adapter 本地连接 2:

    Connection-specific DNS Suffix . :
    IP Address. . . . . : 172.16.2.205
    Subnet Mask . . . . : 255.255.255.0
    Default Gateway . . . . : 172.16.2.254

C:\>ping **172.16.2.1**
Pinging 172.16.2.1 with 32 bytes of data:
Reply from 172.16.2.1: bytes=32 time=1ms TTL=255
Reply from 172.16.2.1: bytes=32 time<1ms TTL=255

S2#**show dot1x all**
Sysauthcontrol   Enabled
Dot1x Protocol Version   3
Dot1x Info for FastEthernet0/10

PAE   = AUTHENTICATOR
QuietPeriod   = 60
ServerTimeout   = 0
SuppTimeout   = 30
ReAuthMax   = 2
MaxReq   = 2
TxPeriod   = 30

S2#**show authentication interface f0/10**

```
Client list:
  Interface    MAC Address         Method    Domain    Status          Session ID
  Fa0/10       54ee.7545.035b      dot1x     DATA      Authz Success   AC10010200000012004FB6D4
Available methods list:
  Handle   Priority   Name
    3         0       dot1x
Runnable methods list:
  Handle   Priority   Name
    3         0       dot1x
```

6.8 实验 7：SPAN

1. 实验目的

通过本实验可以掌握：
① 在交换机上配置 SPAN 和 VSPAN 的方法。
② 在交换机上配置 RSPAN 的方法。

2. 实验拓扑

SPAN 实验拓扑如图 6-32 所示，本实验将在安装有抓包软件的 Test PC 上监测数据包。

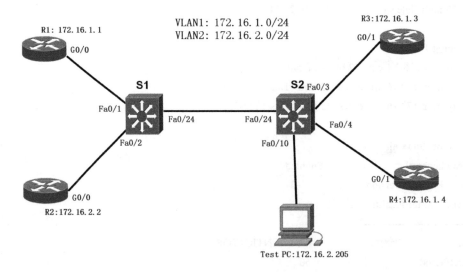

图 6-32　SPAN 实验拓扑

3. 实验步骤

（1）准备工作

在 S1 上配置 VLAN 间路由，保证 R1、R2、R3、R4 和 Test PC 之间可以通信。

```
S1(config)#vlan 2
S1(config)#interface fastethernet0/2
S1(config-if)#switchport mode access
S1(config-if)#switchport access vlan 2
//R2 在 VLAN2 上
S1(config)#interface range fastethernet0/22-23
//关闭这些端口，该实验不需要
S1(config-if-range)#shutdown
S1(config-if)#interface fastethernet0/24
S1(config-if)#switchport trunk encapsulation isl
S1(config-if)#switchport mode trunk
S1(config-if)#switchport trunk native vlan 888
//把 Native VLAN 设为不存在的 VLAN，使得 Trunk 链路上不会存在 Native 帧
S1(config)#ip routing
S1(config)#interface vlan 1
S1(config-if)#no shutdown
S1(config-if)#ip address 172.16.1.254 255.255.255.0
S1(config)#interface vlan 2
S1(config-if)#no shutdown
S1(config-if)#ip address 172.16.2.254 255.255.255.0


S2(config)#vlan 2
S2(config)#interface fastethernet0/10
S2(config-if)#switchport mode access
S2(config-if)#switchport access vlan 2
S2(config)#interface range fastethernet0/22- 23
S2(config-if-range)#shutdown
S2(config)#interface fastethernet0/24
S2(config-if)#switchport trunk encapsulation isl
S2(config-if)#switchport mode trunk
S2(config-if)#switchport trunk native vlan 888


R1(config)#interface GigabitEthernet0/0
R1(config-if)#no shutdown
R1(config-if)#ip address 172.16.1.1 255.255.255.0
R1(config)#ip route 0.0.0.0 0.0.0.0 172.16.1.254


R2(config)#interface GigabitEthernet0/0
R2(config-if)#no shutdown
R2(config-if)#ip address 172.16.2.2 255.255.255.0
R2(config)#ip route 0.0.0.0 0.0.0.0 172.16.2.254


R3(config)#interface GigabitEthernet0/1
R3(config-if)#no shutdown
```

R3(config-if)#**ip address 172.16.1.3 255.255.255.0**
R3(config)#**ip route 0.0.0.0 0.0.0.0 172.16.1.254**

R4(config)#**interface GigabitEthernet0/1**
R4(config-if)#**no shutdown**
R4(config-if)#**ip address 172.16.1.4 255.255.255.0**
R4(config)#**ip route 0.0.0.0 0.0.0.0 172.16.1.254**

测试 R1、R2、R3 和 R4 之间的通信应该是正常的。在 Test PC 上配置 IP 地址 172.16.2.205/24，不配置网关，测试与 R2 之间的通信，应该是正常的。

（2）在 S2 上配置 SPAN

S2(config)#**monitor session 1 source interface fastethernet0/3 both**
//配置 SPAN 的源为 Fa0/3 端口的收、发流量
S2(config)#**monitor session 1 destination interface fastethernet0/10**
//配置 SPAN 的目标端口为 Fa0/10。默认时，一旦 Fa0/10 成为目标端口，Test PC 不能从 Fa0/10 端口发送流量。

在 Test PC 上开启抓包软件 WireShark，然后从 R3 上 ping R4，应该能抓到它们之间的流量，抓包结果如图 6-33 所示。

图 6-33 抓包结果

再次测试 Test PC 与 R2 之间的通信，应该不能通信。当 Fa0/10 端口成为 SPAN 的目的端口后，除非特殊配置，否则该端口将不能接收数据包。

S2(config)# **monitor session 1 source interface fastethernet0/24 both**
//增加 Fa0/24 作为 SPAN 的源，该端口上接 Trunk，封装 ISL

在 Test PC 上开启抓包软件 WireShark，然后从 R1 上 ping R4，应该能抓到它们之间的流量，抓包结果如图 6-34 所示。

图 6-34 抓包结果

从以上抓到的包看出，包并没有进行 ISL 封装。这是因为，默认时，当数据帧从目的端

口发送出时，ISL 封装会被去掉以本帧（Native）发出。可以更改命令为：

S2(config)#**monitor session 1 destination interface fastethernet0/10 encapsulation replicate**
//配置复制到目的端口的帧将保持源端口的封装。复制到目的端口的帧封装和目的端口上的配置无关，即使 Fa0/10 端口上配置了 mode access 或者封装 dot1q，最终 Test PC 收到的还是源端口上帧的格式

再次在 Test PC 上开启抓包软件 WireShark，然后从 R1 上 ping R4，应该能抓到它们之间的流量，抓包结果如图 6-35 所示。注意 ISL 标签的存在。

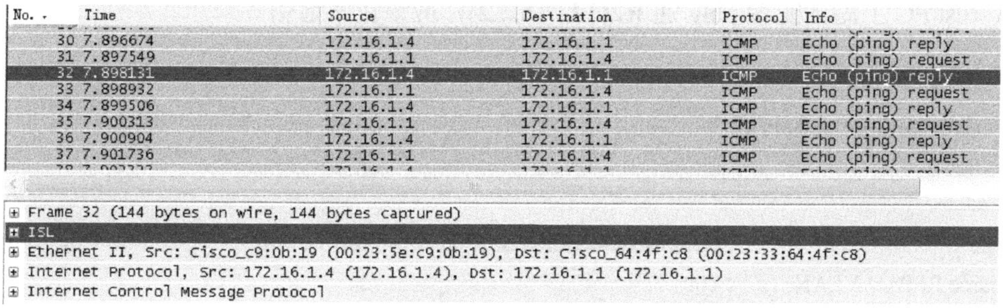

```
No.    Time       Source       Destination    Protocol  Info
 30  7.896674   172.16.1.4    172.16.1.1      ICMP     Echo (ping) reply
 31  7.897549   172.16.1.1    172.16.1.4      ICMP     Echo (ping) request
 32  7.898131   172.16.1.4    172.16.1.1      ICMP     Echo (ping) reply
 33  7.898932   172.16.1.1    172.16.1.4      ICMP     Echo (ping) request
 34  7.899506   172.16.1.4    172.16.1.1      ICMP     Echo (ping) reply
 35  7.900313   172.16.1.1    172.16.1.4      ICMP     Echo (ping) request
 36  7.900904   172.16.1.4    172.16.1.1      ICMP     Echo (ping) reply
 37  7.901736   172.16.1.1    172.16.1.4      ICMP     Echo (ping) request

⊞ Frame 32 (144 bytes on wire, 144 bytes captured)
⊞ ISL
⊞ Ethernet II, Src: Cisco_c9:0b:19 (00:23:5e:c9:0b:19), Dst: Cisco_64:4f:c8 (00:23:33:64:4f:c8)
⊞ Internet Protocol, Src: 172.16.1.4 (172.16.1.4), Dst: 172.16.1.1 (172.16.1.1)
⊞ Internet Control Message Protocol
```

图 6-35　抓包结果

技术要点

注意 monitor session 1 destination interface fastethernet0/10 encapsulation dot1q 或者 monitor session 1 destination interface fastethernet0/10 encapsulation isl 命令。意思是：复制到目的端口的帧将重新封装为 dot1q 或者 isl，然而在 Catalyst 3560 交换机上，该命令实际上无效。

还有一个很重要的问题：WireShark 抓包软件有可能看不到帧中的 dot1q 和 isl 标签，这是因为有的网卡在收到带 dot1q 和 isl 标签的帧后通常会把标签剥离。

（3）配置 RSPAN

S1(config)#**vlan 100**
S1(config-vlan)#**remote-span**
S2(config)#**vlan 100**
S2(config-vlan)#**remote-span**
//创建 VLAN100，作为 RSPAN VLAN

S1(config)#**monitor session 1 source vlan 2 rx**
//配置 SPAN 的源为 VLAN1 的接收流量
S1(config)#**monitor session 1 destination remote vlan 100**
//配置 SPAN 的目的 VLAN 为 VLAN100

S2(config)#**no monitor session 1**
//删除 monitor session 1，因为它使用 Fa0/10 端口
S2(config)#**monitor session 2 source remote vlan 100**
//配置 monitor session 2 的源为 remote vlan 100
S2(config)#**monitor session 2 destination interface fastethernet0/10 ingress vlan 2**
//配置 monitor session 2 的目的端口为 Fa0/10 端口。Ingress 允许 Fa0/10 端口接收 Test PC 发送的帧，并且数据帧以 Native 封装（即没有标签），数据帧发往 VLAN 2

在 Test PC 上开启抓包软件 WireShark，然后从 R1 上 ping R2，应该能抓到它们之间的流量，

抓包结果如图 6-36 所示。注意：只抓到了 VLAN2 的接收流量，即 R2 到 R1 的 ICMP 回包 Echo reply。

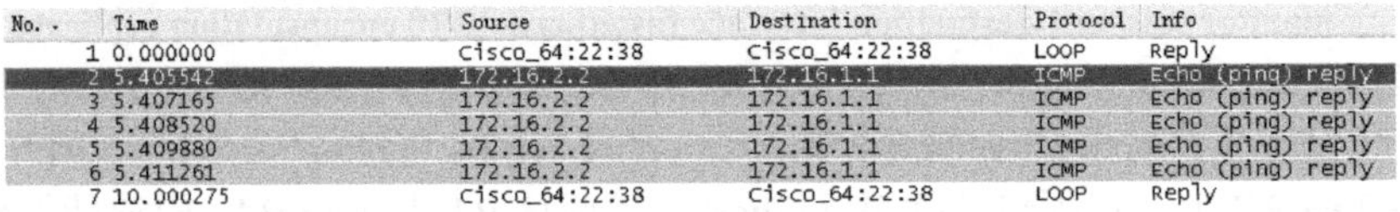

No.	Time	Source	Destination	Protocol	Info
1	0.000000	Cisco_64:22:38	Cisco_64:22:38	LOOP	Reply
2	5.405542	172.16.2.2	172.16.1.1	ICMP	Echo (ping) reply
3	5.407165	172.16.2.2	172.16.1.1	ICMP	Echo (ping) reply
4	5.408520	172.16.2.2	172.16.1.1	ICMP	Echo (ping) reply
5	5.409880	172.16.2.2	172.16.1.1	ICMP	Echo (ping) reply
6	5.411261	172.16.2.2	172.16.1.1	ICMP	Echo (ping) reply
7	10.000275	Cisco_64:22:38	Cisco_64:22:38	LOOP	Reply

图 6-36　抓包结果

从 Test PC 上测试能否 ping 通 R2(172.16.2.2)，应该正常通信。

（4）show 命令

```
S2#show monitor session 2
Session 2
---------
Type                     : Remote Destination Session      //类型
Source RSPAN VLAN        : 100                             //源
Destination Ports        : Fa0/10                          //目的端口
    Encapsulation        : Native                          /帧的封装方式
          Ingress        : Enabled, default VLAN = 2       //目的端口是否允许接收帧
    Ingress encap : Untagged                               //目的端口接收的帧封装格式
```

6.9　实验 8：RACL、VACL 和 MAC ACL

1. 实验目的

通过本实验可以掌握：

① 在交换机二层端口上配置 RACL 和 MAC ACL 的方法。

② 在交换机上配置 VACL 的方法。

2. 实验拓扑

RACL、VACL 和 MAC ACL 实验拓扑如图 6-37 所示。

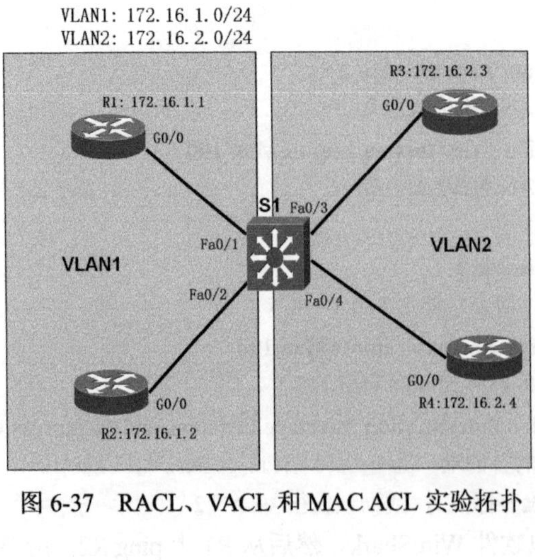

图 6-37　RACL、VACL 和 MAC ACL 实验拓扑

3. 实验步骤

（1）准备工作

在 S1 上配置 VLAN 间路由，保证 R1、R2、R3 和 R4 之间可以通信。

```
S1(config)#vlan 2
S1(config)#interface range fastethernet0/1 - 2
S1(config-if)#switchport mode access
S1(config)#interface range fastethernet0/3 - 4
S1(config-if)#switchport mode access
S1(config-if)#switchport access vlan 2
S1(config)#ip routing
S1(config)#interface vlan 1
S1(config-if)#no shutdown
S1(config-if)#ip address 172.16.1.254 255.255.255.0
S1(config)#interface vlan 2
S1(config-if)#no shutdown
S1(config-if)#ip address 172.16.2.254 255.255.255.0

R1(config)#interface GigabitEthernet0/0
R1(config-if)#no shutdown
R1(config-if)#ip address 172.16.1.1 255.255.255.0
R1(config)#ip route 0.0.0.0 0.0.0.0 172.16.1.254

R2(config)#interface GigabitEthernet0/0
R2(config-if)#no shutdown
R2(config-if)#ip address 172.16.1.2 255.255.255.0
R2(config)#ip route 0.0.0.0 0.0.0.0 172.16.1.254

R3(config)#interface GigabitEthernet0/0
R3(config-if)#no shutdown
R3(config-if)#ip address 172.16.2.3 255.255.255.0
R3(config)#ip route 0.0.0.0 0.0.0.0 172.16.2.254

R4(config)#interface GigabitEthernet0/0
R4(config-if)#no shutdown
R4(config-if)#ip address 172.16.2.4 255.255.255.0
R4(config)#ip route 0.0.0.0 0.0.0.0 172.16.2.254
```

注意：测试 R1、R2、R3、R4 能够互相通信。

（2）配置 MAC ACL

注意：MAC ACL 只对非 IP 流量起作用，MAC ACL 能够根据帧的源和目的 MAC 地址及帧的类型进行过滤。

```
S1(config)#mac access-list extended MACL
```

S1(config-ext-macl)#**deny f872.ead6.0000 0000.0000.ffff any**
//命令格式为{**deny | permit** } *SourceMAC MASK DestMAC MASK EtherType*。拒绝源 MAC 地址为
f872.ead6.****的帧。路由器 R1 的 G0/0 端口的 MAC 地址为 f872.ead6.f4c8

S1(config-ext-macl)#**permit any any**
//定义了一个 MAC ACL

S1(config)#**interface fastethernet0/1**
S1(config-if)#**mac access-group MACL in**
//在 Fa0/1 端口接收方向上应用 MAC ACL

再重新测试 R1 和其他路由器的通信状况，很可能还是正常的，原因在于：MAC ACL 对 ICMP 这样的 IP 数据包并不起作用。在路由器 R1 上，执行"**clear arp-cache**"命令，再重新测试 R1 和其他路由器的通信状况，应该无法通信，原因在于：当清除 ARP 表后，R1 需要先发送 ARP 请求，而该 ARP 请求将被 MAC ACL 丢弃。

（3）配置 RACL

可以在交换机的二层端口应用 IP 标准 ACL 或者扩展 ACL，但只能应用在 IN 方向。

S1(config)#**access-list 100 deny ip host 172.16.2.3 host 172.16.1.2**
S1(config)#**access-list 100 permit ip any any**
//拒绝路由器 R3 和 R2 通信，但可以和其他路由器通信

S1(config)#**interface fastethernet0/3**
S1(config-if)#**switchport**
//明确指明端口是二层端口

S1(config-if)#**ip access-group 100 in**
//在端口的 IN 方向应用 IP 的扩展 ACL。

测试路由器 R2 和 R3 之间的通信状况，应该是不正常的；测试路由器 R3 和 R4 之间的通信状况，应该是正常的。

（4）配置 VACL

VACL 应用在 VLAN 上，没有方向性，进入或离开 VLAN 的数据都要受控制。

S1(config)#**interface fastethernet0/3**
S1(config-if)#**no ip access-group 100 in**
//先清除 Fa0/3 上的 ACL

S1(config)#**access-list 101 permit ip host 172.16.2.4 host 172.16.1.2**
S1(config)#**vlan access-map VACL 5**
S1(config-access-map)#**match ip address 101**
S1(config-access-map)#**action drop**
//配置从 R4 发往 R2 的数据将被丢弃。

S1(config-access-map)#**vlan access-map VACL 10**
S1(config-access-map)#**action forward**
//配置其他的数据包将被转发。

S1(config)#**vlan filter VACL vlan-list 1**

测试路由器 R2 和 R3 之间的通信状况，应该是正常的；测试路由器 R2 和 R4 之间的通

信状况，应该是不正常的。注意：VACL 是应用在 VLAN1 上的，当从 R2 发包到 R4 时，VACL 并不丢弃数据包；而当 R4 发包到 R2 时，VACL 丢弃数据包。

6.10 本章小结

　　本章首先介绍了交换机上的访问安全和端口安全，防止黑客取得交换机的控制权，而交换机的端口安全则是防止 MAC 泛洪攻击的最有效手段。DHCP 嗅探是防止 DHCP 服务器伪装的手段，同时 DHCP 嗅探又是防止 ARP 欺骗和 IP 欺骗很重要的前提。有了 DHCP 嗅探、DAI 和 IPSG，局域网中的 ARP 病毒和源 IP 地址伪装则无藏身之处。VLAN 跳跃攻击的防范则容易得多，禁止端口上的 DTP 即可。AAA 和 dot1x 能够对合法的用户进行授权，可以防止外来用户接入网络。除此之外还可以通过各种访问控制列表，防止非法访问。流量镜像（SPAN）经常用于入侵检测或者入侵保护。

参 考 文 献

[1] [美] 弗鲁姆，[美]西瓦萨布拉玛尼安，[美]弗拉姆. CCNP 学习指南：组建 Cisco 多层交换网络（BCMSN）（第 4 版）. 刘大伟，张芳，译. 北京：人民邮电出版社，2007.

[2] [美] 载伊，[美]麦克唐纳，[美]鲁菲. 思科网络技术学院教程 CCNA Exploration：网络基础知识. 思科系统公司，译. 北京：人民邮电出版社，2009.

[3] [美] Richard Froom，[美]Reum Frahim. CCNP SWITCH 300-115 学习指南. 孙玲，韩鹏，译. 北京：人民邮电出版社，2016.